RDS: The Radio Data System

For a complete listing of the *Artech House Mobile Communications Library,*
turn to the back of this book.

RDS: The Radio Data System

Dietmar Kopitz
Bev Marks

Artech House
Boston • London

Library of Congress Cataloging-in-Publication Data
Kopitz, Dietmar.
    RDS : the radio data system / Dietmar Kopitz, Bev Marks.
       p.   cm.
    Includes bibliographical references and index.
    ISBN 0-89006-744-9 (alk. paper)
       1. RDS (Radio)   I. Marks, Bev.   II. Title.
    TK6570.R27K67   1998
    621.384'152—dc21   98-41083
                                                              CIP

British Library Cataloguing in Publication Data
Kopitz, Dietmar
    RDS : the radio data system
    1. Radio - Packet transmission
    I. Title   II. Marks, Bev
    621.3'845

    ISBN 0-89006-744-9

Cover design by Lynda Fishbourne

© 1999 EBU

International Standard Book Number: 0-89006-744-9
Library of Congress Catalog Card Number: 98-41083

10 9 8 7 6 5 4 3 2

Contents

Foreword

RDS was developed as a result of the far sighted preparatory studies undertaken within EBU Technical Groups over 20 years ago. The system was designed to fulfill the requirements of all European countries and it subsequently became a European standard under the umbrella of CENELEC. In many countries RDS services were rapidly introduced with the aim of generally improving FM radio and especially mobile reception. New data services, in particular for traffic and travel information, were added and are now being introduced. RDS has been further developed to permit migration to Digital Radio which has even more powerful features built upon that experience already gained.

In recent years, the European-developed RDS has also become a global success. In the United States, RDS was first adapted to meet North American requirements, then the RDS Forum, with its worldwide viewpoint, stressed the need for converging standards. This was recently achieved by joint activity in Europe and in the United States, culminating in upgraded, harmonised standards for RDS in Europe and RBDS in the United States.

It is recognised that FM radio will still exist for many years to come. So, in the future the EBU will continue to support the maintenance of the RDS standard. But of course, one day Digital Radio will deliver much more powerful data services, a process started many years before by RDS.

The authors, Dietmar Kopitz and Bev Marks, have accompanied the development of RDS from the very earliest days. They are now highly active in the related domain of increasing importance to broadcasters—the provision of radio data services specifically for traffic and travel information.

I wish this book a great success and I hope that it will also stimulate many new initiatives for further implementations of RDS all around the world.

Professor Albert Scharf

Acknowledgments

In the European Broadcasting Union, many working groups have contributed to the elaboration of the Radio Data System, since the early nineteen seventies. We have had the privilege and pleasure to work with these groups for many years. As a result, we both enjoy long lasting friendships with many highly gifted personalities who have contributed so much to the success story that RDS has already become, with over 50 million RDS radios in use.

Much of the content of this book is based on shared knowledge gained during the multinational development work to which many people from other countries have contributed. We cannot individually mention everyone, however, we would like to list the most significant contributors and express our appreciation to them for their contributions given to RDS. These individuals have made RDS internationally successful through standardisation in Europe. We are also pleased to acknowledge the support of their organisations or companies.

- *Josef Berger (Österreichischer Rundfunk), Austria*
- *Kari Ilmonen (Yleisradio), Finland*
- *Martti Saarelma (Yleisradio), Finland*
- *André Keller (TéléDiffusion de France), France*
- *Michel Rigal (TéléDiffusion de France), France*
- *Philippe Meillan (TéléDiffusion de France), France*
- *Hermann Eden (Institut für Rundfunktechnik), Germany*
- *Jürgen Mielke (Institut für Rundfunktechnik), Germany*

xx RDS: The Radio Data System

- *Karl-Heinz Schwaiger (Institut für Rundfunktechnik), Germany*
- *Mario Cominetti (Radiotelevisione Italiana), Italy*
- *Henri van der Heide (Nederlandse Omroep Stichting), Netherlands*
- *Theo Kamalski (Philips Car Systems), Netherlands*
- *Sten Bergman (Sveriges Radio), Sweden*
- *Tore Karlsson (Televerket), Sweden*
- *Østen Mäkitalo (Televerket), Sweden*
- *Christer Odmalm (Televerket), Sweden*
- *Ernst Schwarz (Swiss PTT), Switzerland*
- *Johnny Beerling (British Broadcasting Corporation), United Kingdom*
- *Stan M. Edwardson (British Broadcasting Corporation), United Kingdom*
- *Bob (S R) Ely (British Broadcasting Corporation), United Kingdom*
- *Simon Parnall (British Broadcasting Corporation), United Kingdom*
- *Mark Saunders (British Broadcasting Corporation), United Kingdom*
- *Ian Collins (UK Independent Radio), United Kingdom*

After the European RDS standard was established within CENELEC, a new standardisation activity started in the US National Radio Systems Committee and an adaptation of RDS to the North American broadcast environment resulted in the agreement of the EIA/NAB voluntary industry standard: RBDS. Again we would like to acknowledge the significant work of additional North American contributors.

- *Terry Beale (Delco Electronics)*
- *John D. Casey (Denon Electronics)*
- *Almon H. Clegg (Denon Nippon Columbia)*
- *Jerry LeBow (Sage Alert Systems)*
- *Thomas D. Mock (Electronic Industry Association)*
- *Dave Wilson (National Association of Broadcasters)*
- *Scott A. Wright (Delco Electronics)*

We thank the EBU for the permission to use, for the purpose of this book, RDS material that has been elaborated over the years in our daily work. We are grateful to Philippe Juttens (EBU), for his intuitive understanding of our needs, resulting in the high quality graphical design work for RDS that we have regularly used in many EBU publications.

Our many contacts with members of the RDS Forum have given us much help and inspiration, for which we are grateful.

We also thank our publisher Artech House and, in particular, Julie Lancashire and John Walker, our editors, for their continuous encouragement to progress this project. We greatly appreciated their guidance during the development of the book concept which we conceived together for the mobile communications series.

We thank our book reviewer, Grant Klein for the excellent professional advice given to us during the writing of the manuscript. Susanna Taggart from Artech House has also given considerable and valuable help to us in this context.

Finally, we thank our families and close friends for their long patience with us. Perhaps inevitably with such a subject, we significantly underestimated the time needed to accomplish the manuscript and they were always kindly forgiving when the writing of the book made us unavailable and temporarily prevented us from enjoying life together. Now that all the hard work is done, we hope that we will find many new opportunities together which will compensate for the good times missed during the winter of 1997!

Dietmar Kopitz and Bev Marks
Geneva (Switzerland) and Battle (England)
October 1998

1

RDS System and Applications Overview

1.1 Introduction

This chapter is conceived to give a good and detailed overview about the Radio Data System (RDS) and its origins, and does not require too much technical knowledge from the reader about mobile data communication techniques. It provides much of the necessary background that will help readers to better understand the details given about RDS and its implementation options in the remainder of this book.

1.2 Objectives to be Achieved With RDS

The Radio Data System offers broadcasters a flexible data transmission channel accompanying their very high frequency/frequency modulation (VHF/FM) sound broadcasts. Additionally, RDS offers the possibility for data service providers to introduce new data services if these are based on the concept of sending relatively few bits to many users. Thus, RDS can accommodate a wide range of possible implementation options.

Following a long period of systems development in the 1970s and early 1980s (see Figure 1.1), and field trials in several European countries, RDS is now implemented all over Western Europe, in several Central and East European countries, in some Asia Pacific region countries, in South Africa, and in the United States (using the Radio Broadcast Data System (RBDS) standard), and is also used by some broadcasters in Latin America. One important new feature for which regular services started in many European countries as of

Figure 1.1 One of the first RDS demonstration receivers designed by the BBC in 1982. (*Source:* BBC.)

1997 is the Traffic Message Channel (TMC), (see Chapter 7). Another important new feature is the Open Data Application (ODA), (see Chapter 9).

1.3 Historical Development

Early in the 1970s, many public broadcasters in Europe were beginning to ask themselves what could be done with FM. It had been introduced in the 1950s and yet it was none too successful, despite continued investment in the transmission infrastructure. Many big broadcasters had, by the mid-1970s, completed their national FM networks with nominal service coverage of around 95% of the population, or more. Nevertheless, audience research and FM receiver sales continued to suggest that something was impeding the take-up of FM radio services by the public. However, in particular, the in-car entertainment sector had worked hard on improving receiver sensitivity, which helped improve reception significantly. Some other factor must have been playing a role in this slow acceptance of FM services. Various research organisations were asked to look at this situation and reported mixed but highly constructive solutions.

In 1974, we had in Europe the following situation: The largest German car radio manufacturer, Bosch/Blaupunkt had developed, in close collaboration with the research institute of the German public broadcasters (IRT), the Autofahrer Rundfunk Information (ARI) System, which means "broadcast information for motorists." The system used the 57 kHz subcarrier with a 3.5 kHz injection level as a means to identify that the so-marked programme carries from time to time announcements about road traffic. This subcarrier was then amplitude-modulated with 125 Hz when the traffic announcement was broadcast as a means of identifying that such an announcement was on-air. In addition, one out of six possible signals (between 23.75 Hz and 53.98 Hz) was used for area identification.

Bosch/Blaupunkt was hopeful at that time that this ingenious system would be adopted by the broadcasters all over Europe, which would have been an advantage from the receiver manufacturer's point of view because of the convenience of a more uniform market for the sale of car radios. To gain the broadcasters' support, the ARI system was submitted by the German public broadcasters to the European Broadcasting Union's technical committee, with the view of obtaining a recommendation from the EBU that this system be put into general use all over Western Europe.

The EBU is a professional association of, at that time, mostly public broadcasters in Western Europe, but now also includes the broadcasters of Central and Eastern Europe. The EBU is in fact the authority to establish or harmonise operational broadcast practises in Europe. In doing so, there is full awareness in the EBU that it is not a standardisation organisation. Therefore, the EBU collaborates very closely with standardisation organisations like the International Telecommunication Union (ITU), Comité Européen de Normalisation Electrotechnique (CENELEC), and European Telecommunications Standards Institute (ETSI) to create the necessary standards, normally before any recommendation relating to an operational practice for broadcasting is issued.

Although it was rather unexpected by those who undertook the initiative in the EBU—to recommend the ARI system for general introduction in 1974—their motion launched the RDS development within the EBU. Why? In the EBU's technical committee there was a great deal of disagreement about the universal applicability of the ARI system. The broadcasting model used in Germany, and for which the ARI system was conceived, was in fact rather exceptional. Instead of regional broadcasting companies, most countries used national networks. Regionalisation, though quite useful for road traffic information, was not a common practise at that time. Also, for ARI it was assumed that in each region there would only be one programme that contained broadcast information for motorists. In reality, though, national broadcasters inserted these announcements in several of their programmes. Thus, within the technical committee of the EBU, in 1975, a number of provoking questions and statements were being put forth, such as the following:

1. Would it not be better to seek to develop a system that uses digital modulation instead of the analogue AM used in ARI?

2. Why should we adopt a system that permits identification of only one programme, namely the one that contains the traffic announcements? It would be much better to develop a universal system that permits identification of any FM programme—for example, by Programme TYpe.

3. The hand-over mechanism for broadcast networks, by means of the area codes used within ARI, is inconvenient from the broadcasters' point of view, since it does not permit identification, unambiguously, of the possible alternative transmitters within a given network; that is, Alternative Frequency lists are required instead.

These criticisms of the ARI system immediately set the scene for the RDS development to start. There was general agreement within the EBU that this would be a very useful undertaking. The task was given to a working group that was in charge of all questions related to sound broadcasting. This group, in fact, took some time to take off the ground, since it had no experience at all with the use of digital modulation systems. Therefore, after having reflected upon the most suitable subcarrier frequency (57 kHz or 76 kHz, both integer multiples of the 19 kHz pilot tone) for the purpose of achieving a minimum of interference, the group started to work on compatibility issues. They covered such aspects as interference from the data signal to the stereophonic audio programme, the required coverage area (the same as for monophonic reception), and the ARI compatibility. Additionally, the aim was to achieve no degradation of the established protection ratios that are internationally used within the ITU for the purpose of frequency planning of broadcast networks, or even single local transmitters.

The EBU working group then created a specialised group of experts in data broadcasting. In most European countries, by the late 1970s, the public broadcasters and the telecom organisations that operated transmitter networks had already experimented with data transmissions where a subcarrier within the FM multiplex signal was phase-modulated. This kind of experience existed especially in Scandinavian counties—for example in Finland and Sweden.

The EBU technical committee had, at that time, a so-called "bureau," which was their small management committee supervising the activities of the associated working groups while also being responsible for organising the work decided on by the full committee. In that bureau there was one member from the Finish broadcasting company, Yleisradio, who had already written his doctoral thesis about the technology that was about to be developed by the EBU's specialist group.

It is interesting to note, even from the present point of view, what Mr. K. Ilmonen's thesis in 1971 was all about, and what kind of research work he had then initiated within the technical department of Yleisradio. One of his collaborators had also joined the EBU specialist group and contributed to the work then being undertaken. Ilmonen's thesis was about listener preferences for loudness in speech and music broadcasts when these occur at various sequences in the same programme. To permit a separate adjustment of the volume and some kind of automatic control function in broadcast operations and the receiver, an identification of each speech or music item was suggested. If this could be done, one could also make an identification of the Programme TYpe. He then drew up a list that closely resembles those lists now used in RDS and DAB. He suggested using a 57 kHz subcarrier, amplitude-modulated by FSK frequencies, to achieve the objective for such a universal identification system. Being in the EBU and the representative of a small country, Ilmonen insisted strongly that Europe needed a standard for a unified system, thus giving a strong impetus on the management level to conduct the work with this very important objective clearly in mind (see the historical document reproduced on page 6) [1].

How did the EBU specialists then proceed in their work? In 1976, there were already several different radio data systems proposed from Finland, the Netherlands (see the historical document reproduced on page 6), and Sweden. The specialists tried to identify what these systems had in common. They looked at a form of coding of the data stream that would permit optimal performance in the mobile reception mode at typical car-travelling speeds and subject to severe multipath interference, as would usually occur with FM in mountainous regions.

To determine these basic parameters, it was agreed to conduct a first field trial in 1980 in the area of Bern/Interlaken, Switzerland. Representatives from the European receiver manufacturing industry (EUROTECH, now the European Association of Consumer Electronics Manufacturers (EACEM)) were invited to join. A questionnaire was sent to broadcasters and industry leaders to determine the desirable features of the upcoming system. The test data broadcast in the region of Bern/Interlaken was then recorded by various research laboratories and analysed with the view of optimising the mobile reception.

In 1981, there was subsequent agreement of coordinated applications and the principles to be used in baseband coding. Test transmissions then started in several countries such as France, Finland, Germany, the Netherlands, Sweden, and the United Kingdom. Since the system parameters were not yet fully defined, each country had designed its own particular radio data system, and sometimes one country even tested several different variants. Thus, by 1982 eight different systems were already known and it became an imminent task to bring the choice straight down to one [2–6].

Two historical EBU documents on emerging RDS

NOS Hilversum, Netherlands:
Some basic proposals for the development of a programme identification system in Band II
On the 14th of October 1975 delegates from IRT Hamburg, NV Philips Eindhoven and NOS were in Hilversum to discuss theproblems concerning the realisation of a transmitter and programme code in the FM band.

A summary of the meeting is given in this document.
....
FM transmitters can be modulated with frequencies up to 100 kHz. Within this frequency range two small bands can be pointed out for extra information. One of them is just around 57 kHz and is used by the Germany ARI system. Unfortunately no extra information can be modulated upon the carrier of 57 kHz.

The second frequency range is around 67 kHz, about 65-69 kHz and is used by the SCA system. This system is not in use in Europe.
....
The transmitter and programme code could be transmitted in a digital form. A practical type of modulation could be the frequency shift keying system, shifting from i.e. 66-68 kHz.

Due to the low modulation depth, the signal to noise ratio within this channel will be fairly low. Probably the system is not suitable for a car or a portable radio.

Information that could be transmitted:

1. Programme identification   4 characters
2. FM channel number   3 characters
3. Place name of the transmitter tower   5 characters
4. Kind of programme   2 characters

Use could be made of the ASCI code. The whole information should be given five times per second. The type of display is still under study.

Fields tests in the near future will give information how far the system could disturb the radio programme and how far the system can be disturbed by man made noise.

                                                                            H.J. v.d. Heide, NOS
                                         ****

Oy. YLEISRADIO Ab.   Helsinki, January 23rd, 1976
The Finnish Broadcasting Company Ltd.

Dear Mr Kopitz,
....
With the present state of art, a radio listener meets several situations of inconvenience in connection with the listening. Firstly, almost everywhere more than one programme are receivable simultaneously on VHF, and it is not always convenient for the listener to know whether he is tuned to the right programme. Secondly, a programme is often transmitted by several VHF transmitters, a number of which are receivable at the listener's place. To find the transmitters with the strongest signal is not easy, would however be useful. Thirdly, listeners waiting for a certain programme item to start (say, the news), frequently wish to ignore the preceding programme, which results in an inconvenience: keeping a constant watch on the run of time. And, finally, research has shown that different programmes are wanted with different characteristics of reproduction, the most widely known differences being that between speech and music, or between music of different type.

By transmission of a suitable programme information signal together with the programme signal proper, all the above inconvenience could be avoided. Receivers could be developed to perform in fact any kind of changes in the listening characteristics: to find a certain programme (including the search for a traffic radio transmitter), to find a programme of a given type (including traffic programmes) to switch on at the right time, to give each programme a given level or a given setting of treble and bass, combination of loudspeakers, etc.
....
For the further development of the programme identification system, we are at present constructing a prototype transmitter and receiver, with the aim of demonstrating to our receiver industry the scope of possibilities such a system could give for the listener-consumer. We have chosen 57 kHz as sub-carrier, amplitude modulated by FSK frequencies. The code chosen allows for transmission of 1024 alternatives of programme identification. The complete data sequence consists of 12 bits.

We have heard that similar developments are going on in research laboratories of some receiver manufacturers. In order to avoid a marketing of receivers with different characteristics, we very strongly consider it of paramount importance to standardise the identification codes. This, according to our opinion, is to a great extent a task for the receiver industry. However, the system (especially if the code is large) will affect the possibility of using e.g. the band 53-75 kHz of VHF transmitters to any other purposes (SCA, tetraphony, the temporary EBU traffic-VHF systems, etc.). Thus, the EBU and CCIR should also investigate the problem.

                                                                            Dr. Kari Ilmonen

The year 1982 saw the EBU specialists defining, prior to any further evaluation, the objective criteria upon which the choice should be made, and they agreed to jointly conduct a laboratory and field test in Stockholm. Figure 1.2 shows the EBU group that developed the RDS system. Out of this evaluation, the Swedish Programme Identification (PI) System emerged as the winner, and was then retained as the basis for further RDS development in the EBU. This PI system was already in use in Sweden, since 1978, for the operation of an FM data broadcasting paging system called MBS [7,8].

Subsequently, an ad hoc group was created to meet at the BBC Research Department with the task of fixing the baseband coding for all known applications to cover with the unified FM Radio Data System.

The features thus coded were tested in a second field trial in the area of Bern/Interlaken (see Figure 1.3). Once the data was evaluated by the research laboratories involved, the RDS specification was drawn up in final form by the EBU specialists meeting in 1983 in Bern, Switzerland.

The European car radio manufacturers who were consulted were still quite concerned about the EBU's RDS system meeting the requirement for ARI compatibility, because that system had, since its introduction in 1974, been very successful in Germany, Austria, Switzerland, and Luxembourg. The majority of all car radios sold in these countries were equipped for ARI functionality.

Other European countries were less interested, and did not use ARI at all. Nevertheless, since RDS was designed to be compatible with ARI, the challenge of successfully passing a field trial had to be attempted to confirm that compatibility to those manufacturers who remained doubtful. Of course, such a field trial had to be carried out in Germany, in an area where mobile reception was as difficult as the one encountered in the Bern/Interlaken area. Munich was chosen for this field trial, which took place in 1983. The RDS successfully passed this rather critical test.

Figure 1.2 The EBU working group that developed the RDS system. (*Source:* BBC.)

Figure 1.3 Basic field trials taking place in the area of Interlaken (Switzerland) in 1982. (*Source:* EBU.)

As a consequence, the RDS specification was adopted by the EBU and EUROTECH in 1983-84, published, and also submitted to the ITU and CCIR Study Group 10. This study group extracted the essential characteristics from the EBU specification and transcribed them to a new CCIR Recommendation 643, which was then adopted in 1986 [9].

All the above shows clearly how RDS has emerged over time and between the years 1975-1984. In retrospect, during this 10-year period we saw the following occur:

1. The desire to universally identify each Frequency Modulation (FM) programme; this created the PI and Programme Service (PS) features.

2. The desire to identify broadcasts for motorists more universally than ARI; this created the Traffic Programme (TP) and Traffic Announcement (TA) features.

3. The desire to hand over a mobile receiver within a network; this created the Alternative Frequencies (AF) feature.

4. The desire to identify speech and music and programme types; this created the Music Speech (MS) and Programme Type (PTY) feature.

5. The desire to maintain radio paging within the data broadcast as it was already implemented in Sweden; this created the Radio Paging (RP) feature.

A system was now designed and available for the mobile listener, who had needed help with in-car reception of FM for the various reasons already established by audience research; namely, automatic retuning from one transmission coverage area to the next area and so on. The system also provided an emulation of the ARI system used in Austria, Germany, Luxembourg, and Switzerland that alerted drivers to traffic announcements. Many other features were proposed and built into the RDS system to be dynamically multiplexed as needed in each transmission. The key mechanisms were designed for mobile reception and a group/block data format to ensure very fast data synchronisation and decoding of certain features, while allowing some features to be conveyed at a slow rate for general information. Since the system design was developed by broadcasters working in the well-regulated environment of the 1970s, a number of features were considered but not fully developed at that time. However, their far-reaching decisions regarding future enhancements has allowed RDS to mature over the years.

In 1985, EUROTECH agreed with the EBU about the general introduction of RDS and promised, on the condition that the EBU would give their support towards the development of the RDS-Traffic Message Channel (RDS-TMC) feature (see Chapter 7), that the first RDS receivers would be presented at the international consumer electronics show, IFA'87, in Berlin. From 1988 on, these receivers would be marketed in all those countries where RDS was already introduced.

Given the fact that the RDS development was so well coordinated by the EBU, broadcasters in all European countries, through this activity, were fully aware of the benefits created for their listeners (some said they could now surf the radio waves). The introduction of RDS throughout Europe happened fast—so fast, indeed, that some then called it the "silent revolution" [10–15]. Figure 1.4 illustrates the high FM spectrum occupation which makes RDS necessary.

Broadcasters started to implement RDS transmissions with a mixture of self built RDS encoders, which led to the growth of a small, specialised professional equipment market selling RDS encoders and associated RDS monitoring equipment. The earliest implementations were undertaken by some large network broadcasters, and they selected just a few RDS features to start their trials and preservice activities. Within a couple of years, some problems had come to light as these initial transmissions began to give evidence that the original standard was somewhat lacking when real-world situations were faced.

In 1988, the BBC officially launched RDS. Johnny Beerling, chairman of the EBU Programme RDS Experts Group and cochairman of the RDS Forum, remains a strong supporter of RDS. Figure 1.5 shows Beerling at the BBC opening ceremony.

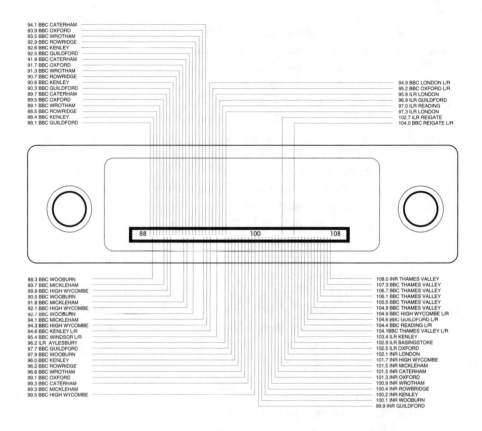

Figure 1.4 Why RDS was needed is clearly demonstrated by the high FM spectrum occupation, as this example from London shows. (*Source:* EBU.)

In 1988–89, when receivers were ready to conquer the European market, RDS was already on-air over most of Western Europe.

Table 1.1 displays the historical development of RDS.

1.4 Evolution of the RDS Standards

1.4.1 Europe

1.4.1.1 EBU

The first RDS standard was published in March 1984. It was titled "EBU Tech 3244: Specifications of the Radio Data System RDS for VHF/FM sound

Figure 1.5   Johnny Beerling at the BBC opening ceremony in 1988. (*Source:* BBC.)

Table 1.1
History of RDS Development

1975	Pre-development start
1980	First field trial at Bern/Interlaken, Switzerland
1982	Test start in Stockholm, Sweden
	Evaluation of eight systems in Helsinki, Finland
	RDS baseband coding agreed
	Second field trial at Bern/Interlaken
1983	Industry/broadcasters meeting at EBU
	Joint industry/broadcaster field trial in Munich, Germany
	RDS adopted by EBU and industry—submitted to CCIR
1984	First presentation of RDS in Detroit, MI, USA
	Ford starts RDS car radio development in Detroit
	RDS specification EBU 3244 published
1985	Large scale pre-operational trial in Germany
	EBU recommends RDS introduction
	Industry/broadcasters agree first receivers target from 1987
1986	First presentation of RDS at NAB Dallas, TX, US
	RDS CCIR Recommendation published
1987	Ireland, France and Sweden introduce RDS
	First RDS receivers shown at IFA Berlin, Germany
	Volvo markets the world's first RDS car radio

Table 1.1 (continued)

1988	Austria , Belgium, Denmark, Germany, Italy and the United Kingdom introduce RDS Blaupunkt, Grundig and Philips mass produce RDS car radios
1989	RDS enhancements: EON developed and tested in the UK Norway, The Netherlands, Portugal and Switzerland introduce RDS Presentation of RDS in Washington DC and NAB Las Vegas, NV, US
1990	First presentation of RDS at BroadcastAsia in Singapore and in South Africa CENELEC adopts RDS as the European standard EN 50067
1991	First RDS-EON receivers shown at IFA Berlin First presentation of RDS in China RDS presentation in New Orleans, LA to US Public Radio Hong Kong introduces RDS
1992	New version of CENELEC RDS standard published South Africa introduces RDS USA: EIA/NAB RBDS standard completed which includes RDS
1993	RDS Forum created to promote RDS implementation Grundig: presents at IFA Berlin first portable RDS receiver
1994	European Commission recommends RDS-TMC for Trans-European Road Network First European DGPS implementation in Sweden Universal encoder communication protocol enhanced
1995	RDS Paging Association created EIA activates RDS promotion in the United States First RDS Forum meeting in the United States
1996	RDS Forum enhances RDS CENELEC standard NRSC in the United States agrees with RDS Forum to harmonize RBDS and RDS
1997	New RDS CENELEC standard submitted to vote New RBDS NAB/EIA US voluntary standard submitted to vote UECP enhanced to conform with new RDS CENELEC standard Germany - first country to introduce RDS-TMC

broadcasting," and it contained some 14 different RDS features, as shown in Table 1.2 [27].

It is very illuminating to realise how the publication of a very technical and specific niche standard (notice how it was called a "specification") can affect all of us, as consumers, for evermore. It is calculated that there are now some 50 million RDS receivers worldwide in the hands of consumers by the end of 1997, and a very high proportion of those use abbreviations like AF, TA, and TP, for example, on their front panels or in their displays. Did the standards writers realise the impact their work would have? These abbreviations we

Table 1.2
List of RDS Features Defined in the Original EBU Specification in 1984

EBU Tech 3244 – Features	Abbreviation
List of Alternative Frequencies	AF
Clock Time and date	CT
Decoder Identification	DI
In-House applications	IH
Music Speech switch	MS
Information concerning Other Networks	ON
Programme Identification	PI
Programme Item Number	PIN
Programme Service Name	PS
Programme TYpe	PTY
RadioText	RT
Traffic-Announcement identification	TA
Transparent Data Channel	TDC
Traffic-Programme identification	TP

now live with do not appear to be very user friendly, and most consumers have been subjected to them not knowing their origins. This is indeed a lesson for all designers to consider very carefully for future broadcast systems: The laboratory quick-fit naming solutions need careful consideration for long-term user friendliness. However, the RDS designers were indeed very far sighted technically, as we shall see later.

It is true to say that up until 1984, not too many receiver designers had considered RDS, because this standard came from the research laboratories of broadcasters and not from commercial receiver manufacturers. But that situation changed, and the commercial receiver companies soon realised the benefits that RDS had been designed to bring to the broadcasters' listeners and to their future customers. Within a year, development work was being undertaken in both Europe and elsewhere, and the first RDS receiver came from a car company, Volvo. Volvo was anxious to improve car safety through the introduction of several automatic features that an RDS receiver could provide. Figure 1.6 depicts Volvo's worldwide commercial RDS car radio in 1987. Of course, almost all well-known commercial car receiver companies now produce RDS

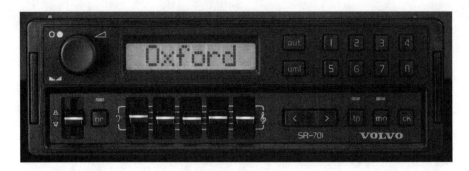

Figure 1.6 In 1987, the Volvo 701 first commercial RDS car radio in the world. (*Source:* Volvo.)

receivers, and this came about because the broadcast sector was also committed to RDS and started to introduce RDS transmissions across Europe.

Between 1984 and 1989, four supplements to the original specification were issued, covering the following areas:

- Alternative Frequencies: methods A and B;
- Radio Paging;
- Programme TYpe code definitions;
- Enhanced Other Networks.

With the perspective of that era, there is no doubt that the EON development was a major change to the standard, which had come about from the joint efforts of broadcasters and receiver designers attempting to implement a system that allowed signalling from one network belonging to a broadcaster to another network of the same broadcaster. Experience had shown that the ON mechanism of the original standard just did not work!

After much thought, at an EBU meeting held in July 1987 and a number of subsequent meetings to distill the details, several new concepts were developed, including EON, which could give a receiver a full "picture" of a broadcaster's networks over a two-minute period. Then, dynamic signalling could vector a receiver to specific services as needed; for example, a travel bulletin could be received from another transmission in the area of reception. By common consent, BBC Radio agreed to become the field test site for these techniques, and implemented EON during 1988 with signalling associated to five local radio services and referenced by the BBC Radio networks. That trial was very successful and the United Kingdom became a continuing test site for many RDS receiver designers from all over the world. These designers came to test their software implementations of EON for second-generation RDS receivers.

Over the years, RDS has also attracted a number of different RDS encoder manufacturers. Originally, each chose a communications protocol for use between studio and transmitter site where the RDS encoders are installed to achieve dynamic control of the transmitted RDS data. Initially, this aspect of RDS escaped the standardised approach to RDS, perhaps because the manufacturers efficiently satisfied their client broadcasters and very few initially requested dynamic control. But gradually, the need for features such as TA flag control—and more recently, the PTY feature and RadioText, which have recently been more and more frequently implemented in RDS receivers—have shown that broadcasters definitely need dynamic control. What is more, because over time they have wished to purchase RDS encoders from several sources, the need for a standardised protocol became evident.

Under the auspices of the EBU, many major RDS encoder manufacturers have cooperated to develop the Universal Encoder Communications Protocol (UECP, see Chapter 11), which is now managed and maintained by the RDS Forum. During 1997, it reached version 5.1, published as SPB 490. This protocol, now briefly called the UECP, allows broadcasters to specify associated network servers and RDS control systems that will use a common data format, which will then enable easy installation with existing RDS encoders. Over time, the UECP was gradually upgraded as the transmission standard required. Normally, this has been carried out approximately yearly, with the information and the UECP specification circulated to all members of the RDS Forum, since the RDS encoder manufacturers are nearly all members and the overhead costs of a public standard would be too great for this very small niche market to sustain [16].

1.4.1.2 CENELEC

With significant development work going on in European industrial manufacturing companies, it became clear that a better recognised standard would serve well to publicise the RDS system and ensure consistent design activity in the diverse organisations working on RDS products. So, in 1988, the technical committee TC 107 (now TC 207) of the European Committee for Electrotechnical Standardisation, CENELEC, began (in close cooperation with the EBU) to transcribe EBU Tech 3244 and the four supplements mentioned above into a European standard, EN 50067. This was published in December 1990, and took over as the definitive standard for Europe.

EN 50067, published in 1990, became the "solid rock" that the RDS system needed, both from a broadcaster's standpoint to ensure reliable transmissions and also from the RDS receiver designer's point of view. Both these parties required the RDS standard to link them together and give each the certainty that RDS would indeed give the radio user the assistance that RDS had promised nearly 10 years before.

So, here was the first significant demonstration that the original development of the RDS specification had been given future-proofing, as the additional features in the four supplements could be added quite easily and allow continued development of both the transmission equipment and RDS receivers.

Nevertheless, standards also require stability to allow development time, and the issue of EN 50067:1990, with endorsement from both the broadcast and the manufacturing sectors, promised this time in the future. But, of course, Europe could not now keep RDS to itself. Already some broadcasters from other parts of the world had noticed what RDS could do. Notably, broadcasters in two very diverse countries—Hong Kong and South Africa—started to negotiate for RDS technology. They were prepared to invest in RDS to eventually promote the technology to both their customers (the listeners) and to the RDS receiver suppliers. In the absence of a worldwide standard for RDS they naturally opted for RDS implementations using EN 50067:1990, especially because the consumer manufacturers could only offer products manufactured for that standard. In both of these cases, the broadcasters had similar structures to those already found in Europe, so the RDS standard requirements fit their networks well and virtually no adaptation was forced upon them.

If a standard has been well designed, additions to it will offer enhancements that the industry and the consumer alike will want. However, the timing of such enhancements has to be considered carefully to ensure that the revised standard does not destabilise the marketplace. Accordingly, the EBU issued SPB 482, which proposed certain enhancements to EN 50067:1990. These enhancements were made to clarify, to a greater level of detail, certain coding issues that had become necessary. That work was ready for the next issue of the standard and completed well in advance of the marketplace actually needing the items that were standardised. This was to prove valuable in the developments that were taking place in the United States. In effect, parallel discussion proceeded over the next 24 months or so, and EN 50067:1992 was issued in April of that year. It just missed the work of another small EBU group who had developed PI codes and extended country code (ECC) proposals for the worldwide implementation of RDS, and their output was published in August 1992 by the EBU as SPB 485 (revised in 1992) covering allocation of country/area identification codes in RDS [17,18].

In the meantime, CENELEC EN 50067:1992 has been much upgraded. The new revised text, which was first published in September 1996 by CENELEC, was prepared by the RDS Forum in close collaboration with the

CENELEC technical committee 207 with full involvement of experts from the EBU. Certain elements of text were revised in accordance with experience gained from the RDS system and changes in broadcasting practise since the initial specification was published. An interesting example are the new clauses relating to the PS feature.

The Open Data Application (ODA) has been added as a new feature to permit a flexible extension of RDS to still undefined applications. Furthermore, cross-references were made to the Comité Européen de Normalisation (CEN) standards, defining the RDS-TMC feature.

Receivers produced in accordance with the new specification will, of course, be compatible with RDS broadcasts, which conform to previous editions of the RDS specification.

The resulting new European standard, EN 50067:1998, replaced from April 1998 onwards the old EN 50067:1992, and all earlier versions as well. The differences between the new and the old standard versions are summarised in Table 1.3 [19–22]. Table 1.4 lists the RDS features in upgraded RDS/RBDS standards.

Table 1.3
Comparison Between the New and the Old CENELEC RDS Standards

New: CENELEC EN 50067:1998			Old: CENELEC EN 50067:1992	
Section	Additions/Modifications	Page	Section	Page
0. Scope	Extended	6	0. Scope	6
1. Modulation characteristics of the data channel	Unchanged	6–11	1. Modulation characteristics of the data channel	6–11
2. Baseband coding	Unchanged	12–14	2. Baseband coding	12–14
3. Message format	PS usage explicitly restricted to programme service label	15–53	3. Message format	15–49
	ODA added			
	TMC standard references added			
	Static/Dynamic PTY flag added			
	PTYN added			
	RP transferred to Annex M			
4. Description of features	Extended	54–57	4. Glossary of terms of applications	50–53
5. Marking	Unchanged	57	5. Marking	54

Table 1.3 (continued)

New: CENELEC EN 50067:1998			Old: CENELEC EN 50067:1992	
Section	Additions/Modifications	Page	Section	Page
Annex-A: offset words	Extended	59	Annex-A: offset words	55
Annex B: modified shortened cyclic code	Unchanged	60–65	Annex B: modified shortened cyclic code	56–61
Annex C: group and block synchronisation	Unchanged	66–68	Annex C: group and block synchronisation	62–64
Annex D: PI and ECC codes	Updated	69–72	Annex D: PI and ECC codes	65–70
Annex E: character definitions	Introduction modified Character tables unchanged	73–76	Annex E: character definitions	71–74
Annex F: PTY codes	Modification and 15 new codes added	77–80	Annex F: PTY codes	75–77
Annex G: time and date conversions	Unchanged	81–82	Annex G: time and date conversions	78–79
Annex H: ARI system	Unchanged	83	Annex H: ARI system	80
Annex J: language identification	New	84–85		
Annex K: RDS logo	RDS-EON logos added	86	Annex K: RDS logo	82
Annex L: Open Data registration	New	87–89		
Annex M: Radio Paging	Extended with enhanced paging	90–125		
Annex N: world-wide country codes	New	126–130		
Annex P: abbreviations	Updated	131	Annex J: abbreviations	81
Annex Q: bibliography		132	Annex L bibliography	83

Table 1.4

List of Possible Defined RDS Features in New Upgraded RDS/RBDS Standards

Code	Name	Function	Description
PI	Programme Identification	Tuning	16-bit code giving a unique serial number to a programme service
PS	Programme Service name	Tuning	You see what you hear
			An 8-character label of the programme service, e.g. "Radio 21"
AF	Alternative Frequencies	Tuning	List of other frequencies carrying the same or a related programme service
TP	Traffic Programme identification	Tuning	1-bit flag indicating of whether this programme service will carry traffic announcements
TA	Traffic Announcement identification	Switching	An on/off signal to indicate when a traffic announcement is on air
PTY	Programme TYpe identification	Radio programme related	Identifies from a list of 31 possibilities the current programme type, e.g. "News" or "Sport"
PTYN	Programme TYpe Name	Radio programme related	8-character label to indicate a more specific programme type, e.g. PTY = 4: "Sport", PTYN: "Football"
EON	Enhanced Other Network information	Tuning / switching	Provides a cross-reference to PI, PS, AF, PTY for vectorised switching to other programme services, usually operated by the same broadcaster
CT	Clock Time and date	Time reference	Universal time code with local offset and modified Julian date
RT	RadioText	Radio programme related	Text of max. 64 characters for display
PIN	Programme Item Number	Radio programme related	Scheduled start time and date for an individual programme
M/S	Music Speech code	Radio programme related	1-bit flag to indicate music or speech, e.g. for separate volume controls
DI	Decoder Identification	Radio programme related	4-bit code to indicate mono, stereo and other possibilities
TMC	Traffic Message Channel	Radio data service	Coded predetermined traffic messages for output in various languages

Table 1.4 (continued)

Code	Name	Function	Description
ODA	Open Data Application	Radio data service	Open or encrypted service permitting to implement any possible function as given RDS capacity would permit
IH	In-House applications	Radio data service	Data which are used by the broadcaster or network operator
TDC	Transparent Data Channel	Radio data service	Unformatted text or data in 32 possible channels
RP	Basic Radio Paging and enhanced paging	Radio data service	Paging service using FM broadcasts as the transport mechanism
EWS	Emergency Warning System	Radio data service	Provides for the coding of warning messages
ECC	Extended Country Code	Tuning	Extends the 16 possible PI country codes to a unique country identifier
	Linkage	Tuning	A flag with a PI reference carried within EON of another programme service to which the tuned programme will be or is already linked

1.4.1.3 CEN

In the mid-1980s, the EBU RDS experts were prompted by European car radio manufacturers, and EUROTECH in particular, to consider an RDS feature that was quickly given the name Traffic Message Channel (TMC). Indeed, by the time EN 50067:1992 was published, TMC had the RDS group type 8A allocated. This feature was soon recognised by traffic management experts in Europe as a potentially very valuable feature since it permitted the delivery of coded traffic messages which, in-car, could be interpreted in a driver's native language regardless of the visited country. This was a simple idea but, unfortunately, many complex issues were associated with this feature and the European Commission has funded much research into the application.

In this process, many new standards were developed about messages, dictionaries for all the languages needed, and the management issues concerning these elements. The coding of RDS-TMC has been undertaken by many workers, coordinated by CEN TC 278. Field trials in the mid-1990s showed that RDS-TMC will work well, but significant infrastructure requirements will be needed to implement RDS-TMC fully across Europe, and this phase is now just beginning. Chapter 7 deals with all these issues in much more detail.

1.4.2 United States

In 1990, discussions started about standardising RDS for the United States under the auspices of the National Association of Broadcasters (NAB), and the National Radio Standards Committee (NRSC) was asked to form a subcommittee to report on the possibilities.

In the U.S. radio environment, radio networks and relay transmitters, or transposers as they are called in the United States, are more infrequently found, which is quite different from Europe. So, RDS clearly needed some adaptation. But the NRSC subcommittee, which had elected to call the American standard "Radio Broadcast Data System (RBDS)," realised that RDS would be more quickly implemented in the United States if core aspects of both systems were shared, because RDS knowledge, RDS encoders, and RDS receivers were all readily available. Indeed, the subcommittee that worked on RBDS standardisation even wished for as much harmonisation as they could achieve.

Apart from a few new features, the RBDS standard required special interpretation of two of the existing features. Firstly, the PI code structure of EN 50067:1990 was unsuited to the different regulation of radio stations in the United States, where "call letters" are the only centralised data that can be relied upon. Thus, a clever algorithm was developed to allow conversion of call letters into a unique PI code so that existing RDS encoders and RDS receivers could use this PI code without any problem. The reason for this approach was that one wanted to avoid the need for a federal organisation to be charged with administering PI codes for RBDS implementation. Secondly, the U.S.-specific programme format of radio stations needed a new list of PTY codes since PTY codes for Europe were quite inappropriate. Thus, a different set of PTY codes was developed for the United States. The other significant demonstration was that RDS could largely be compatibly upgraded as time progressed and new ideas were required.

Generally, it was thought that RDS receivers could now be used anywhere in the world, provided the ECC feature was used to uniquely designate a country. The original RDS specification had only considered countries that were members of the EBU; now, expansion to the whole world became a distinct necessity.

In the United States, the RBDS specification was adopted in January 1993 as a voluntary national standard, jointly issued by the EIA and the NAB. As explained above, this standard includes as its major component the RDS system, and European receivers could easily be modified for use in the United States. In the large majority of cases, they would even work well unmodified, especially with the five basic features: PI, PS, AF, TP and TA [23].

The differences between RBDS and RDS are explained in greater detail in Chapter 2. Simultaneously, as the European RDS standard was upgraded in the years 1995–97, an upgraded RBDS specification was completed by the end of 1997 within the National Radio Systems Committee (NRSC), which is jointly sponsored by the Electronics Industry Association/Consumer Electronics Manufacturers Association (EIA/CEMA) and the National Association of Broadcasters (NAB) [24].

RBDS was now drawn up with the view to harmonise, to the largest extent possible, the RBDS specification with the RDS features most recently specified in Europe in the EN 500067:1998.

In the United States since about 1995, RBDS is only the name for the American standard. When implemented in receivers, the system has been called RDS, as it is anywhere else in the world. RDS in the United States is identified with the same logo as specified in the European CENELEC standard, which is the same as was once developed for the EBU by the BBC.

1.5 System Maintenance and Promotion

While successful standards are in use by the few people who helped develop them, all is likely to be well because they know what they intended when drafting the documents! But once a wider group of users has a need for a standard, the intentions that are not specified fully or well, can be misunderstood or misinterpreted. Furthermore, field experience of implementing standards tends to throw up many new issues.

Recognising a need for more information about RDS, the EBU published the document, "Tech 3260, Guidelines for the implementation of the RDS system," which was intended to encapsulate and transfer some of that knowledge to other workers in RDS technology before it was lost [25].

With all the developments that have taken place over recent years, it is now necessary to prepare another edition of the RDS guidelines document, and the RDS Forum proposes to undertake that work with the support of many workers in the RDS field. It will have a slightly differently focus as to implementations and the system approach needed to achieve a full set of RDS services, including non-programme-related data services multiplexed with radio programme-related data services, and the commercial and technical issues associated with managing and billing in such a system.

1.5.1   RDS Forum: a Worldwide Association of RDS Users

The RDS Forum has existed since 1993. Membership is open to all professionals involved in using RDS technology. The RDS Forum has held two plenary meetings per year, and a large proportion of the more than 100 members worldwide attend.

In 1997, the RDS Forum had four working groups concerned with maintaining the RDS standard, developing accepted guidelines for RDS system operation, upgrading the Universal Encoder Communications Protocol (UECP), and dealing with RDS/DAB cross-referencing together with the objective of implementing plans for DAB transmissions and receivers that offer a compatible user interface.

The operational expenses of the association are shared among all the membership. Members pay an annual fee for each registered representative. More detailed information about the RDS Forum is available on the Internet.

The Internet address for the RDS Forum Web site is: www.rds.org.uk.

1.5.2   The United States: NAB and EIA/CEMA

RDS is promoted in the United States through the National Association of Broadcasters (NAB) and the Consumer Electronic Manufacturers Association (CEMA), a branch of the Electronics Industry Association (EIA). Both the NAB and CEMA organise very significant annual conventions, with a major technical exhibition where RDS technology is presented at various occasions. The EBU was also invited to participate in these presentations. In addition, both the NAB and CEMA participate in the RDS Forum, which is coordinated by the EBU. Hence, there is a continuous exchange of experiences with regard to the implementation of RDS in Europe and the United States.

CEMA sponsored an RDS promotional effort in 1995–96. While there are some 5,000 FM radio stations operating in the United States, after that campaign, 750 of them had RDS implemented. The campaign concentrated on the 26 metropolitan areas with the largest population in the United States and was aimed at raising the awareness of radio broadcasters about RDS. At the same time, the NAB published a 98-page booklet explaining RDS applications to broadcasters [26].

One particular problem that existed at that time in the United States was that there were not many RDS receivers on the market, and therefore dealers and consumers were largely unaware of RDS and what benefits it could offer. In addition, many broadcast station owners were confused about RDS because the NAB was of the opinion that the United States needed a high-speed data system (HSDS) on a subcarrier around 76 kHz, in addition to RDS. Many broadcasters understood that this was because RDS had too many limitations (670 usable bits per second versus 10-16 kbps) with respect to the number of additional data services that it could support. They also thought that once agreement was reached about a high-speed data system, this would make RDS redundant. But this is a misperception, since the HSDS subcommittee of the National Radio Systems Committee (NRSC) is seeking a system that is complementary to RDS, in full recognition of the RBDS voluntary industry standard. Figure 1.7 depicts the RDS promotional event sponsored in 1989 by the NAB. This major U.S. convention is where the EBU helped to promote RDS in the United States.

Promotion of the RDS technology in the United States is further initiated through the RDS Advisory Group, a kind of counterpart to the European RDS Forum but without the task to coordinate the further development of the RDS technology. The RDS Advisory Group is an institution of CEMA. Membership is open to all manufacturers, broadcasters, and data service providers interested in the RDS technology. There is no membership fee to be paid and

Figure 1.7 NAB'89 (*Source:* EBU.)

meetings take place during major conventions (on average, four times a year). A quarterly newsletter distributed through CEMA and NAB provides updates on new developments and products in the U.S. market. The activity is financed through voluntary contributions from interested consumer electronic manufacturers. The companies that were most active in the years 1995–97 were Delco, Denon, and Pioneer.

The following Web sites provide more information and updates about RDS in the United States:

- CEMA: www.cemacity.org/rds/
- NAB: www.nab.org./SciTech/

1.6 Usage of RDS Worldwide

1.6.1 Europe

RDS is widely implemented in Europe. Details of the features used can be found in Appendix J.

At the end of 1997, the situation was generally as follows. The basic features PI, PS, and AF (and when applicable, TP and TA) are used by almost all networks and local stations in all countries. Initially, broadcasters using networks had problems implementing radio programme-dependent RDS features, one reason being that there were generally no data links between the studios and the RDS encoders on the transmitter sites that could be used to transport the information.

Nowadays, this situation is changing. Audio links between the broadcasting house and the transmitters have become increasingly digital, and on the radio programme side, the new possibilities of digital radio and radio on the Internet require the radio programme makers to turn their interest towards the future and use the new multimedia. This creates ideal conditions for RDS to have dynamic PTY and RadioText (RT) implemented. The same is also true for MS.

RDS-TMC, a new feature in 1997, is now being more widely used. The driving force behind this is the transport sector of the European Commission and the European Conference of the Ministers of Transport (ECMT). Details about its implementation are given in Chapter 7.

Radio paging (RP) via RDS is used in a number of European countries, but the increasing success of the Global System for Mobile Communications (GSM) mobile telephone services make this less and less interesting. France is still the country where RDS Radio Paging (RP) is most important, but other pockets of RP can be found elsewhere and are still developing.

The Emergency Warning System (EWS) was considered in a number of European countries, but a number of alternative technologies are already in place, so this feature is generally not used. In the United States, there is a much greater interest in using this feature.

The Differential Global Positioning System (DGPS) via ODA is being considered in a number of countries, and these services are already or will soon be operational in Germany, Sweden, and the United Kingdom. However, the system specifications being used are not the same. In Germany, the service will be open, and in the other countries, it is encrypted and a subscription is required by the end user.

1.6.2   The Special Case of Central and East European Countries

The RDS standard CENELEC EN 50067:1998 is restricted to FM Band II (87.5–108 MHz), but in Central and Eastern Europe, FM Band I (65–74 MHz) is also in use. In these countries, FM broadcasting occurs now in both bands; formerly, it was in band I only. In principle, RDS would be feasible in Band I as well, but the necessary AF codes are not defined—neither for Band I nor for any cross-referencing between bands I and II.

A decision was made in the RDS Forum not to provide further support for development of RDS for Band I. This was done to encourage the transfer of radio services operating in Band I to Band II, with the view of harmonising the use of band II for Eastern Europe in the same way as it is used in Western Europe. It was also recommended by the RDS Forum that broadcasting information for motorists should only be implemented in Band II so that RDS-TMC could then become available everywhere in Europe within the same frequency band.

RDS is already used in a number of Central and Eastern European countries, but nowhere yet is it generally implemented on all band II FM stations or networks. Often, the reason for using RDS is the desire to operate an RDS paging service and/or a subscription service for DGPS; where this is the case, the basic RDS features required for tuning, such as PI, PS, and AF, are also implemented.

1.6.3   United States/Canada/Mexico

The people involved in the making of the RBDS standard for the United States generally hold the view that this standard will apply to the whole of North America, the reason being that this is one and the same market for technology.

Officially, this is not yet confirmed. The RBDS standard was made by the U.S. NRSC with this objective in mind, and it is indeed very likely that the market will, in fact, react in this way.

As already described above, only 15% of all U.S. radio stations used RDS by the end of 1997. The main reason was that up until now, there were only a few RDS receivers in the retail shops and dealers and consumers were largely unaware of benefits that RDS could offer to them. This situation is likely to change, and the RDS Advisory Group is active for just this very simple reason. Now, there is greater awareness among the car makers in America about the benefits that RDS can offer to a car radio, and many 1998/99 car models will be equipped with RDS radios for the first time.

RDS-TMC is being tested in some areas of the United States (Arizona, Minneapolis/St. Paul, and a few others), but there is no decision yet to use this technology because it is also thought that a high-speed data system (HSDS) may offer superior technical possibilities. The only drawback is that there are no system standards yet adopted, but the NRSC and also ISO TC 204 remain attentive and are prepared to do some standardisation work on this issue.

1.6.4 Other Countries

RDS is being used in a large number of countries all over the world, and usually these countries use the European CENELEC standard since a world standard for RDS does not yet exist. However, the CENELEC standard already has a large number of hooks that permit countries outside Europe to use this standard without modifying it. For example, the EN 50067:1998 contains an annex with the Extended Country Codes for all countries in the world (see Appendix G). In addition, the character tables specified in one of the annexes support many non-European languages. In certain cases, however, new character tables need to be developed. In such a case, the CENELEC standard is not entirely applicable and needs some modification. Some of the countries concerned are already considering creating a special national standard for themselves. This is the case in China, India, and South Korea.

So far, Japan is likely to remain among those countries where RDS will not be implemented. There are a number of reasons for this. The FM band has a different frequency range, so the AF section of the RDS standard cannot be applied without modification. Also, FM broadcasting doesn't generally have the same importance for radio listeners there as in other countries in the world, because a lot of radio listening still takes place on AM. In addition, there is the belief that because Japan has adopted the data radio channel (DARC) standard (which is compatible with RDS and could well be used in parallel), RDS is completely unnecessary there. This position is particularly true for the Japanese telecommunication authorities. However, disproving this belief is the Japanese consumer electronics industry, which makes and continues to make RDS receivers by the millions for the rest of the world. Thus, one could well imagine that these Japanese manufacturers would be quite happy for RDS to also be used in their own country. The real reason that it is not, is a kind of internal policy issue, which certainly is of no interest whatsoever to the rest of the world.

1.7 System Characteristics

1.7.1 Choice of Modulation Parameters

RDS development started with a number of functional requirements that had to be fulfilled. These are as follows:

1. The radio data signals must be compatible; they must not cause interference to the reception of sound programme signals on existing receivers or to the operation of receivers that use the ARI system.

2. The data signals must be capable of being reliably received within a coverage area as great as that of the monophonic main programme signal.

3. The usable data rate provided by the data channel should support the basic requirements of station and Programme Identification, and provide scope for future developments.

4. The message format should be flexible to allow the message content to be tailored to meet the needs of individual broadcasters at any given time.

5. The system should be capable of being reliably received on low-cost receivers.

These requirements have significantly influenced the choice of the modulation parameters and the baseband coding characteristics.

Figure 1.8 depicts the multiplex spectrum of a stereophonic FM broadcast signal with the small signal level RDS signal centred around the 57 kHz subcarrier, which is the third harmonic of the 19 kHz pilot tone of the stereophonic modulation system. This choice of subcarrier was critical for meeting the requirement to minimise data signal interference to the audio channels for existing receivers. The other parameter that is critical to achieve the same goal is

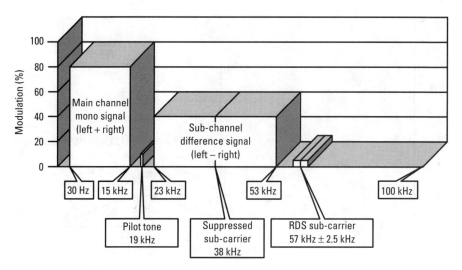

Diagrammatic representation of the baseband signal or "multiplex"

Figure 1.8 Spectrum of a pilot-tone stereo multiplex signal with RDS. (*Source:* EBU.)

the injection level of the data. The higher it is, the more rugged the data service is; however, under multipath conditions, the interference to the audio channels will also increase. It was found in field trials that a minimum was ±1 kHz, and a reasonable operational choice was ±2 kHz. At these levels, there usually is virtually no detectable interference from the data channel during radio listening.

Figure 1.9 depicts the RDS data spectrum in greater detail. As you can see, the spectrum is shaped so as not to interfere with the ARI signals. These use the 57 kHz subcarrier and some additional low frequencies close to the ARI subcarrier to provide amplitude-modulated status and area signalling. As already stated, RDS uses a modulation scheme that was initially developed for the Swedish MBS paging system and that uses a double sideband amplitude-modulated suppressed 57 kHz subcarrier and biphase (sometimes also called "Manchester") coding of the data. The effect of such coding and modulation is to produce a notched spectrum, as is shown in Figure 1.9, so that ARI remains largely unaffected by RDS. Therefore, the two systems can easily coexist.

The use of the biphase-coded data signal also helps compatibility with the audio programme signal because coherent components at around 57 kHz were found to introduce data-modulated crosstalk in receivers that used a phase-locked loop (PLL) stereo decoder, which is the demodulation technique generally used nowadays.

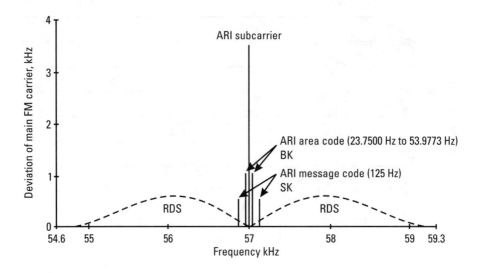

Figure 1.9 Spectra of RDS and ARI signals. (*Source:* EBU.)

The bit rate of the RDS data stream is 1187.5 bits/s (1187.5 = 57,000 / 48) which, with biphase coding and the specified 100% cosine roll-off filtering, gives an overall bandwidth for the data signal of approximately 5 kHz, centred on 57 kHz.

1.7.2 Choice of Baseband Coding

Multipath in an FM system produces distortion of the demodulated signal. The distortion components resulting from the relatively large amplitude sound programme signal components can easily swamp the data signal. When a vehicle moves along a road characterised by multipath interference, the quality of the received FM signal varies rapidly. At some moments, the demodulated audio programme is distorted; at others, it is completely broken up. The very important lesson learned from the 1980 and 1982 field trials in the Bern/Inter-laken area was that reliable mobile reception is only possible when the radio data message stream is broken up into small independent entities, each of which can be received, decoded, and applied independently of other parts of the data stream. This factor was crucial to the basic design of the RDS system and must be clearly understood for the design of new applications within RDS, such as those that can be carried within the ODA feature [14].

As with many other serial data transmission systems designed for mobile communications, the data stream in RDS had to be partitioned into data groups and blocks. How this is done is depicted in Figure 1.10. The groups

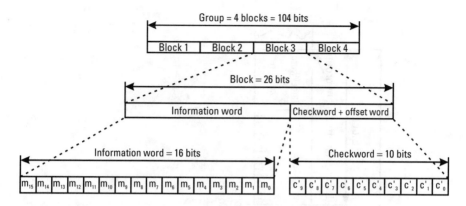

Figure 1.10 Structure of baseband coding of RDS. (*Source:* EBU.)

consist of four blocks, each being 26 bits long. The group thus consists of 104 bits. One block consists of a 16-bit long information word and a 10-bit CRC check word to which is added an offset word that creates the flywheel synchronisation mechanism that makes RDS so rugged under severe multipath receiving conditions.

1.7.3 Message Format and Addressing

Figure 1.11 clearly shows that the RDS coding is structured so that the messages repeated most frequently (and which need a short acquisition time), normally occupy the same fixed positions within a group. This allows decoding without reference to any block outside that contain the information. The first block of each group always contains a PI code, whereas the PTY code and the TP flag occupy fixed positions in block 2 of every group.

The group type code is specified by a 4-bit code that defines the group type (from 0 to 15). This code is sent in the first four bits of the second block of every group. In addition, the fifth bit of this block defines the "version" (A or B) of the group type. In version A groups, the PI code is inserted in block 1 only. In version B, the PI code is inserted in blocks 1 and 3.

Groups are, in general, reserved for a particular application or message type; for example, RT, CT, TMC, and so on.

1.8 Applications of RDS

The Radio Data System permits more than eighteen functions to be implemented. The five most important ones (also called the basic RDS features) are

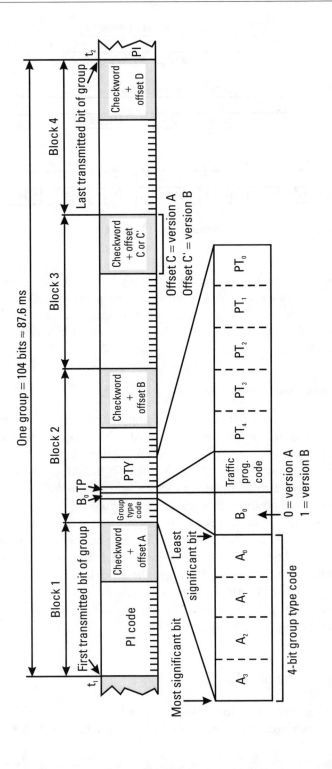

Figure 1.11 Message format and addressing. (*Source:* EBU.)

implemented everywhere and are intended primarily to be used in the mobile reception mode with car radios having automated tuning functions:

- *Programme Identification (PI):* A 16-bit code containing a country symbol, a regional code, and a number permitting the identification of the broadcaster and the particular programme.

- *Programme Service (PS) name:* Comprising eight alphanumeric characters from an uppercase and lowercase character set defined by the EBU, and serving to give the listener information about the name of the programme.

- *Alternative Frequency (AF) lists:* One or more lists can be transmitted, each containing a maximum of 25 frequencies (represented by the corresponding band II channel numbers) of transmitters or rebroadcast transmitters relaying the same programme.

If the programme will contain announcements about local and/or regional traffic situations, the following features should be used to identify this:

- *Traffic Programme (TP) code:* Serving to identify programmes that, from time to time, carry messages addressed to motorists.

- *Traffic Announcement (TA) signal:* Switches a traffic announcement to a preset volume level and, if the motorist is listening to a cassette rather than the radio, stops the cassette and switches the radio on to receive the traffic message instead.

Most of the following other features are optional:

- *Decoder Information (DI):* Indicates one of a number of operating modes for the receiver (e.g., mono, stereo).

- *Music Speech (MS):* A flag indicating whether music or speech is being broadcasted. May be used by the receiver to give independent volume settings of music or speech.

- *Programme TYpe (PTY):* An identification code to be transmitted with each programme item and which is intended to specify the current Programme Type within 31 possibilities. This code could be used for search tuning.

- *Programme Item Number (PIN):* A code identifying a particular programme (by start time and date) to enable automatic on/off switching of receivers.

- *Radio Text (RT):* Either 32 or 64 characters of text for display by receivers.

- *Enhanced Other Networks information (EON):* RDS information relating to other broadcast services. The information includes the PI and AF for quick retuning to the service referenced, TP, TA, PTY, and PIN of these services.

- *Transparent Data Channel (TDC):* Provides for a continuous data stream to receivers and associated peripherals (e.g., printer or home computer).

- *In-House (IH):* A data channel for use only by the broadcaster.

- *Clock Time and date (CT):* A code, usually originated from standard time transmissions, to enable receivers to display the current time and date. This is also used to synchronise various receivers.

- *Radio Paging (RP) and enhanced Radio Paging:* Offers the possibility of mobile pocket pagers with alphanumeric display of messages and alerting bleeps.

- *Emergency Warning System (EWS):* A feature using a very small amount of data for emergency warning services such as national disasters and hazardous chemical spills.

An additional and quite universal feature most recently being developed is:

- *Open Data Application (ODA):* which permits new applications to be designed and implemented in still available data groups with an application identification registration from the EBU or NAB. Interesting examples are differential GPS such as the Rasant system and the control of variable message signs.

1.9 Data Capacity Impact on Applications

Attention is drawn to the fact that the RDS data transmission capacity is rather limited. The system can accommodate only 11.4 data groups per second; that is, only 673 usable information bits per second. This takes into account that each information word contains 16 bits per block and that the data group has five address bits that are used for the group type identification.

Under no circumstances can one implement all possible RDS features within the same data channel. For a better understanding of these constraints, see Appendix E.

1.10   System Performance and Reliability

Here, we briefly summarise what we consider in greater detail in Appendix C. Field tests carried out by a number of broadcasters' research laboratories all came to the same conclusions.

When signals of the ARI system are broadcast simultaneously with RDS, most first-generation RDS demodulators show some significant degradation of the data transmission reliability, which is due to interference from the ARI signals under multipath receiving conditions. This may be overcome however, by filtering out the ARI signal by means of a notch filter at the RDS demodulator input and an appropriate RDS signal injection level of at least ±2 kHz.

One of the methods used to investigate the reliability of the RDS channel for the transmission of real applications such as PI and PS, which are very important for the operation of the automated tuning function of an RDS receiver, is to measure the "waiting time" between successful acquisitions of a particular RDS message.

Relatively low RDS data injection levels, say ±1 kHz, offer a reliable data system only under receiving conditions with little or no multipath effects (typically towns with flat buildings and flat countryside). In a moving receiver, once multipath effects occur due to reflections of the transmitted signal caused by high-rise buildings or mountains, there is a sharp decrease in the reliability for a correct reception of the applications. It all depends on whether the data is repeated sufficiently often in the data stream. In ODA applications, additional CRC check words may also be considered to better protect the data transmission application that is implemented.

Studies usually confirm the ruggedness of the fixed format PI codes compared, for example, with the variable addressed format of the PS codes. Therefore, consumer receivers often store the PS name, displaying the stored name once the PI code is received. Therefore, the use of the PS name to convey, for example, some limited dynamic text information composed of scrolling text or short words of a maximum of eight characters to the radio listener, is just not admissible.

In cases where Radio Paging is implemented within the RDS data stream, the small antenna used in these special paging receivers and the reliability of reception required (which is often within buildings) make an RDS injection level of at least ±4 kHz necessary.

Repetition of message elements transmitted within RDS is also a general requirement. This applies to, for example, RP and TMC messages, but to a lesser degree to RT where the occasional reception of a wrong character will be perceived as less annoying to the reader.

Error detection has to be applied to all messages, and error correction can only be applied to some applications; for example, RT—that is, when an error caused by the correction system is not perceived as being critical.

References

[1]   Ilmonen, K., Listener preferences for Loudness balance of broadcasts with special considerations to listening in noise, Yleisradio (Finnish Broadcasting Corporation), Research Report no. 12/1971.

[2]   van der Heide, H. J., "Possibilities for the introduction and application of a programme code in sound broadcasting," Berichte der 4, NTG-Hörrundfunktagung, Düsseldorf, Germany, Nov. 1976.

[3]   Sundin, L., and A. Sanfridsson, Programme Identification in FM broadcasting, Swedish Telecommunications Administration, Report RI 4022/77, 1977.

[4]   MacEwan, D., "Radio in the 80's Broadcasting and the ideal sound receiver of the future," Wireless World, May 1977, pp. 36–40.

[5]   Ely, S. R., VHF Radio Data: Experimental BBC transmissions, BBC Research Department Report No. BBC RD, 1981-1984.

[6]   Ely, S. R., "The impact of radio-data on broadcast receivers," The Radio and Electronic Engineer, Vol. 52, No. 5, 1982, pp. 291–296.

[7]   Swedish Telecommunication Administration: Paging receiver for the Radio Data System, Doc. 1301/A694 3798 (Alternative B), 1986.

[8]   CCIR: Report 900-1 Radio-paging systems-standardisation of code and format (Annex II), 1986.

[9]   ITU-R Recommendation B.S.643-2 (1995), System for automatic tuning and other applications in FM radio receivers for use with the pilot-tone system.

[10] Mielke, J., and K. H. Schwaiger, "Progress with the RDS system and experimental results," EBU Review, No. 217, June 1986.

[11] Shute, S. A., "The EBU Radio Data System," International Broadcast Engineer, May/June 1987.

[12] Kopitz, D., "Radio Data System-from specification to practical reality," EBU Technical Review, No. 233, Feb. 1989.

[13] Parnall, S. J., and J. L. Riley, "RDS developments," *International Broadcasting Convention Digest*, 1990, pp. 234–240.

[14] Ely, S. R., and D. Jeffrey, *Traffic Information Broadcasting in Mobile Information Systems*, J. Walker, Ed., Norwood, MA: Artech House, 1990, Chapter 5.

[15] Borras Amoedo, J. M., *El RDS: Una Radio Intelligente*, Biblioteca Tecnica PIONEER, Ediciones Tecnicas Rede, S. A., 1991.

[16] Universal Encoder Communications Protocol UECP-EBU doc. SPB 490, version 5.1, European Broadcasting Union, 17A Ancienne Route, CH-1218 Geneva, Aug. 1997.

[17] EBU: Proposed enhancements of the EBU on CENELEC EN 50067 (RDS), Doc. SPB 482, European Broadcasting Union, 17A Ancienne Route, CH-1218 Geneva, Switzerland, 1990.

[18] EBU SPB 485: Rev 1992-Allocation of country/area identification codes in RDS, now included in CENELEC EN 50067:1998.

[19] EN 50067:1990-Specifications of the Radio Data System RDS, European Committee for Electrical Standardisation (CENELEC), 35B rue de Stassart, B-1050 Brussels.

[20] EN 50067:1992-Specifications of the Radio Data System RDS, European Committee for Electrical Standardisation (CENELEC), 35B rue de Stassart, B-1050 Brussels.

[21] Marks, B., "The development of RDS standards," 3rd *Montreux International Radio Symposium: Engineering Symposium Record*, 1996, pp. 133–142.

[22] CENELEC EN 50067:1998 - Specifications of the Radio Data System (RDS.) for VHF/FM broadcasting, European Committee for Electrical Standardisation (CENELEC), 35B rue de Stassart, B-1050 Brussels, Switzerland, April 1998.

[23] EIA/NAB, National Radio Systems Committee: United States RBDS Standard-Specification of the Radio Broadcast Data System (RBDS), Jan. 8, 1993.

[24] EIA/NAB, National Radio Systems Committee: United States RBDS Standard version 2.0, Specification of the Radio Broadcast Data System (RBDS), 1998.

[25] EBU Guidelines for the implementation of the RDS system, Doc. Tech 3260, European Broadcasting Union, 17A Ancienne Route, CH-1218 Geneva, Switzerland, Jan. 1990.

[26] NAB RDS Applications-Opportunities for Radio Broadcasters, NAB Science and Technology Department, 1771 N Street, N.W., Washington D C , 1995.

[27] EBU, Specifications of the radio data system RDS for VHF/FM sound broadcasting, Doc. Tech 3244 and Supplements 1 to 4. European Broadcasting Union, 17A Ancienne Route, CH-1218 Geneva, Switzerland, 1984.

2

Differences Between RDS and RBDS

2.1 Introduction

In this chapter, the small but subtle differences are explained between the European RDS CENELEC standard and the North American RBDS National Association of Broadcasters/Electronic Industry Association (NAB/EIA) specification (a voluntary U.S. industry standard). Through close collaboration between the RDS Forum and the U.S. National Radio Systems Committee, both standards were harmonised to a high degree in order to facilitate the manufacturing of receivers for both markets. Since receivers can detect automatically where they are operated, they can, in principle, configure themselves automatically. Therefore, it is important that encoder/receiver designers and data service providers clearly understand the subtle differences between RDS and RBDS.

2.2 The RDS Component Within RBDS

The RDS for FM broadcasting in the frequency range 87.5–108.0 MHz was developed within the European Broadcasting Union (EBU), as explained in Chapter 1. This system was specified by the EBU in 1984, and has now been introduced in the large majority of European countries as well as a number of other countries all over the world. It is also now being used by a significant portion of the FM broadcast stations in the United States. Later, the system was enhanced through several modifications as more experience was gained during different implementations—in 1987, on large broadcast networks that were then just starting and in 1990, when RDS was adopted as a European standard of CENELEC (EN 50067:1990) [1]. Today, most FM radio stations in Western Europe and several FM radio stations in Central and Eastern Europe use RDS. Receivers—mostly car radios with RDS functionality—are available in Europe from some 75 different manufacturers at prices that are only slightly higher than those of conventional radios.

Although the RDS system offers a wide range of implementation options (some, like the TMC feature, just started implementation in 1997/98), most of the existing RDS car radios have only the five basic features: PI, PS name, AF list, TP identification, and TA signal. PTY code (see Chapter 4) will definitely be one of the popular features in the future, apart from the EON function (see Chapter 6), which is increasingly used now by the European broadcasting networks, and which has led to a second-generation of RDS receivers.

In the United States, a subcommittee on Radio Broadcast Data Systems of the National Radio Systems Committee (NRSC), sponsored by the Electronics Industry Association (EIA) and the National Association of Broadcasters (NAB), began its work in February 1990, after the EBU had demonstrated RDS at various NAB and SAE conventions. The NRSC based its work on RDS as specified within the EBU, and every attempt was made to keep the U.S. standard compatible with it. However, it soon became evident that the completely different broadcasting structure of the United States required that a number of modifications be made. Finally, the U.S. standard had to cover FM broadcasting, as is the case in Europe, but it also had to have certain AM broadcasting-related extensions.

In January 1993, the EIA and the NAB jointly published the U.S. Standard RBDS as a voluntary industry standard that was adopted through a voting procedure that involved both organisations [2].

Since the introduction of the RBDS standard in North America in 1993, much activity has surrounded the introduction of the system by both broadcasters and receiver manufacturers. The EIA created an advisory group to promote the technology to broadcasters and manufacturers, and the RBDS subcommittee remains active for the purpose of upgrading the standard as implementation experience brings about necessary changes. This is completely in line with what was done in Europe only six years earlier. Figure 2.1 shows the EIA booth at the 1993 Chicago Consumer Electronics Show.

Large coordination activities that surrounded the upgrade of the standard took place with the RDS Forum in Europe during the years 1994–1998, with the goal of achieving even more harmonisation between RDS and RBDS. The first result of this harmonisation was that the advisory group proposed the use, at the consumer level, of only one name for the system—namely, only RDS. RBDS is kept as the name for the North American standard to keep it distinct

Figure 2.1 RBDS promotion during the 1993 Chicago Consumer Electronics Show. From left to right: Casey (Denon), Haber (Radio World), Lebow (Sage Alert Systems), and Kopitz. (*Source:* EBU.)

from the European RDS CENELEC standard, since there is no single world standard combining the elements of both standards.

Recent changes to both standards, which were officially adopted in early 1998, have resulted in a significant harmonisation of features, allowing the development of global broadcast equipment, consumer receivers, and data services. The newly revised RDS and RBDS standards both move forward into a new era for FM data services. All changes have been devised to be fully backward compatible with the prior standards while offering a tremendous amount for future growth and extension. The new ODA feature (see Chapter 9) will even permit the definition of additional functionality without the need to revise the RDS/RBDS standards yet again.

Figure 2.2 depicts the actual construction of the RBDS standard. As is shown in this functional diagram of the RBDS standard, the RDS standard is entirely contained within the RBDS standard, and this is true word by word; that is, the corresponding text is identical with the copyright being agreed to by all the organisations concerned. In the RBDS standard, the few remaining functional differences are added within the appropriate sections of the text or within entirely new annexes. This reflects the fact that the functional core of RBDS is completely identical with RDS. This can also be seen by examining the RDS/RBDS group allocations shown in Table 2.1 [3,4].

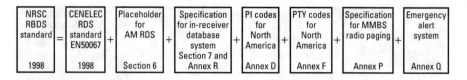

Figure 2.2 Functional diagram of the RBDS standard. (*Source:* NAB/EIA.)

Table 2.1
RDS/RBDS Data Groups

Group Type	Description of Usage
0A	Basic tuning and switching information only
0B	Basic tuning and switching information only
1A	Programme Item Number and slow labelling codes
1B	Programme Item Number
2A	RadioText only
2B	RadioText only
3A	Applications IDentification for ODA only
3B	Open Data Applications—ODA
4A	Clock Time and date only
4B	Open Data Application—sODA
5A	Transparent Data Channels (32 channels) or ODA
5B	Transparent Data Channels (32 channels) or ODA
6A	In-House applications or ODA
6B	In-House applications or ODA
7A	Radio Paging or ODA
7B	Open Data Application—sODA
8A	Traffic Message Channel or ODA
8B	Open Data Application—sODA
9A	Emergency Warning System or ODA
9B	Open Data Application—sODA
10A	Programme TYpe Name
10B	Open Data Application—sODA
11A	Open Data Application—sODA
11B	Open Data Application—sODA
12A	Open Data Application—sODA
12B	Open Data Application—sODA

Table 2.1 (continued)

Group Type	Description of Usage
13A	Enhanced Radio Paging or ODA
13B	Open Data Applications—ODA
14A	Enhanced Other Networks information only
14B	Enhanced Other Networks information only
15A	Undefined (fast PS of RBDS being phased out)
15B	Fast switching information only

2.3 Details About the Differences

2.3.1 Programme TYpe (PTY) Definitions

PTY information allows a user to seek for his or her favourite programme format, such as a particular type of music. Since broadcasting in the United States differs in content, RBDS uses a different Programme TYpe list, as compared to CENELEC RDS. RBDS defines 26 different PTY codes, leaving five codes still unassigned for future extensions, while CENELEC RDS now defines all 32 codes (see Table 2.2).

Table 2.2
European (CENELEC) and North American (US NRSC) PTY Codes

PTY Decimal No.	RDS PTY (CENELEC)		RBDS PTY (US NRSC)	
Number	Description	8-character	Description	8-character
0	No programme type or undefined	None	No program type	None
1	News	News	News	News
2	Current affairs	Affairs	Information	Inform
3	Information	Info	Sports	Sports
4	Sport	Sport	Talk	Talk
5	Education	Educate	Rock	Rock
6	Drama	Drama	Classic rock	Cls_Rock
7	Culture	Culture	Adult hits	Adlt_Hit

Table 2.2 (continued)

PTY Decimal No.	RDS PTY (CENELEC)		RBDS PTY (US NRSC)	
Number	Description	8-character	Description	8-character
8	Science	Science	Soft Rock	Soft_Rck
9	Varied	Varied	Top 40	Top_40
10	Pop music	Pop_M	Country	Country
11	Rock music	Rock_M	Oldies	Oldies
12	Easy listening music	Easy_M	Soft	Soft
13	Light classical	Light_M	Nostalgia	Nostalga
14	Serious classical	Classics	Jazz	Jazz
15	Other music	Other_M	Classical	Classicl
16	Weather	Weather	Rhythm and blues	R_&_B
17	Finance	Finance	Soft rhythm and blues	Soft_R&B
18	Children's programmes	Children	Foreign language	Language
19	Social Affairs	Social	Religious music	Rel_Musc
20	Religion	Religion	Religious talk	REL_TALK
21	Phone In	Phone_In	Personality	Persnlty
22	Travel	Travel	Public	Public
23	Leisure	Leisure	College	College
24	Jazz music	Jazz	Unassigned	
25	Country music	Country	Unassigned	
26	National music	National	Unassigned	
27	Oldies music	Oldies	Unassigned	
28	Folk music	Folk_M	Unassigned	
29	Documentary	Document	Wheather	Wheather
30	Alarm test	TEST	Emergency test	Test
31	Alarm	Alarm_!	Emergency	ALERT_!

Note: "_" indicates a space.

Broadcasters in the United States are very sensitive to the Programme TYpe Name shown on the receiver, because their whole market offering is based on the careful definition of the material they offer. Therefore, Programme TYpe Name (PTYN) Group Type 10A, was already defined in the 1993 version of the

RBDS standard, but it is now included in both the new RDS and RBDS standards. The default PTY eight-character name should be shown during the scanning process. Once the receiver stops on a station with the default PTY category, a more specific PTY category may be given with PTYN; for example, "SPORTS" is coded by PTY and then the sport is detailed as "BASEBALL" by PTYN. The PTYN is not intended to change the default eight characters that will be used during the search, but only to more clearly define the Programme TYpe once tuned to a programme. However, if the broadcaster is satisfied with the default PTY name, he or she will not *need* to use data transmission capacity to add PTYN detail to the default name.

Receivers that implement the PTY feature should allow the user to select one entry from the two different PTY tables (European CENELEC version and North American RBDS version). The table may also be automatically switched using the Extended Country Code (ECC).

2.3.2 Programme Identification (PI) Coding

In North America (United States, Canada and Mexico), PI codes are based on call letters rather than being assigned by a regulatory organisation—as carried out throughout the rest of the world. A portion of the PI codes are reserved for network usage and also for assignment to stations in Canada and Mexico. During the upgrade to the RBDS standard, several mistakes were discovered in the noncall-based PI codes, which had to be corrected. For further details, see Appendix F.

The ECC table was modified for these changes as well. The new PI assignments allow 3,584 possible nonregional PI codes for Canada and Mexico, as well as 765 national network PI codes for all three countries: the United States, Canada, and Mexico.

The ECC codes allow receivers to identify the country that the broadcast is coming from. Since PI Country Identifier codes are limited in number (i.e., there are only 15 codes available), they must be repeated throughout the world. When the PI country identifier code is received in conjunction with the ECC, the exact country of origin can be identified. The updated ECC code table has been expanded to be international in scope, and reflects recent changes in the political landscape. It is contained in Annexes D and N of the RDS and RBDS standards (for details about these codes, see Appendix G in this book). The ECC should be transmitted by all broadcasters, and it is recommended that it be a default automatic transmission in encoders.

The ECC supports the development of global receivers that can automatically compensate for things such as the following:

1. E blocks (MMBS);

2. PI codes;

3. PTY tables;

4. Tuning range and step;

5. FM de-emphasis (North America uses a 75 μs pre-emphasis in the transmitted AF response, while Europe uses 50 μs).

2.3.3 Programme Service (PS) Name

During the update to the CENELEC RDS standard, a movement was made in the RDS Forum to specifically prohibit stations from dynamically changing the PS name [5]. In some countries, broadcasters have used the PS display as a text-messaging feature similar to RT, but only eight characters at a time.

The design intent of PS is to provide a label for the receiver preset that is invariant, since receivers incorporating the AF will switch from one frequency to another in following the selected programme. To combat unintended dynamic use of the PS, the RDS standard adopted strict wording on the use of the PS feature.

During adoption of the RBDS draft, considerable debate about "dynamic" PS resulted in less stringent requirements for PS usage. Set makers should realise this difference—although in actual practical system design they should not prepare for dynamic PS, but just cope with occasional changes of PS as required by the broadcaster to signal a different broadcast network configuration caused by linking or regionalisation.

Excerpts from the text of both standards follows:

- *RDS standard:* The Programme Service name comprises eight characters, intended for static display on a receiver. It is the primary aid to listeners in programme service identification and selection. The use of PS to transmit text other than a single eight-character name is not permitted. Transmission of a PS name usually takes four type 0A groups, but to allow an instant display of the PS when a receiver pre-set is selected, the PS name is often stored for subsequent recall from memory when a programme service is selected. For this reason, PS should generally be invariant. If a broadcaster wishes to transmit longer Programme Service names, programme-related information or any other text, then RT provides this feature.

A similar effect could be experienced with a dynamic text transmission of PTYN. As a result, dynamic PS and PTYN transmissions are expressly forbidden.

- *RBDS standard:* The Programme Service name comprises eight characters. It is the primary aid to listeners in programme service identification and selection. The Programme Service name is to be used only to identify the station or station programme. This text may be changed as required by the station, but shall not be scrolled or flashed or altered in a manner that would be disturbing or distracting to the viewer (i.e., not more frequently than once per minute).

2.3.4 Fast PS Acquisition: Phased Out

The prior version of the RBDS standard included a fast PS feature contained in type 15A groups. This use will now be phased out; however, existing transmissions may still occur for several years. Newly designed equipment should not contain this feature. The type 15A group will remain undefined until at least the next upgrade of the RDS/RBDS standards, which is unlikely to occur before the year 2003.

2.3.5 Optional Multiplexing of RDS and MMBS: Offset Word E

Since 1993, RBDS has included an option for the multiplexing of RDS with the slightly modified MBS (MMBS) system, used exclusively by the company called Cue Paging. MBS is a kind of predecessor to RDS, and MMBS still thrives throughout North America. MBS/MMBS is mainly used as a paging system through a network of approximately 500 stations within the United States. The MBS/MMBS system utilises the same modulation and data structure as RDS but employs a different data protocol. An MMBS broadcast is identified through the offset word E. Since there are similarities between the two systems, it is possible to time division multiplex MMBS and RDS data. Receivers should be able to differentiate between RDS and MMBS broadcasts, but usually disregard the MMBS data, which is only used by specific receivers licensed by the Cue Paging company.

The RDS and MMBS multiplexing will unavoidably degrade the performance for stations that desire to make use of the AF feature due to the loss of repetition of the fast, pertinent tuning information needed by the receiver. A typical time division multiplex of RDS and MMBS data is shown in Figure 2.3. Currently, Cue Paging utilises the MBS (Mobile Search) paging protocol [6] in about 500 stations out of the approximately 5,000 existing FM stations—usually one per area, covering 90% of the population of the United States.

The option of multiplexing between RDS and MMBS will thus permit these stations to implement RDS while maintaining compatibility with the existing MBS paging receivers.

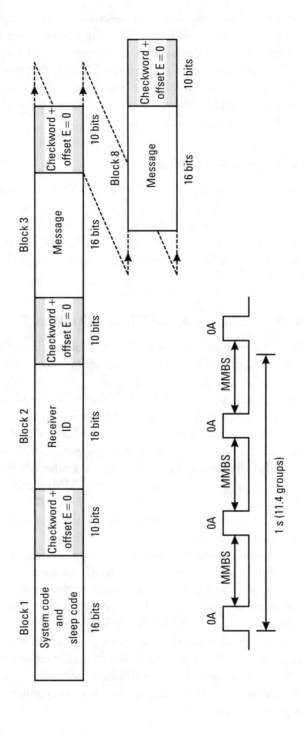

Figure 2.3 RDS/MMBS multiplexing. (*Source:* EBU.)

2.3.6 Optional ID Logic Feature

The ID Logic feature is a licensed technology (PRS Corporation) that allows the incorporation of an in-receiver database that contains the programme format type and call letters for all AM and FM stations. When combined with RDS, ID Logic can provide similar data and features for non-RDS FM and all AM stations. The I-RDS feature allows the database to be updated through an RDS open data application (ODA) so that station information can be updated and maintained automatically. The specification of ID Logic is contained within a separate section of the RBDS standard, while the I-RDS ODA is described in detail within an annex.

According to our knowledge, the ID-Logic feature was not used anywhere in 1997, which is probably due to the difficulty of creating the necessary agreements for a reliable and future-proof broadcast infrastructure to be operated using this technology.

2.3.7 Optional ODA Emergency Alert System

Within the RBDS standard, the emergency alert system ODA protocol is defined for use in the United States. This optional feature set is constructed around the Federal Communication Commission's (FCC) newly developed EAS system and is open for public use. RDS and RBDS allow the silent transmission of emergency information. This has been combined with existing consumer-oriented emergency features to allow additional feature functionality to consumer receivers, as well. The EAS ODA can well accommodate private emergency systems, but it should also be noted that Annex Q of the new RBDS standard has no corresponding section in the RDS standard.

In Europe, there is not yet much interest in this kind of application, and it is thought that it has little or no relevance for consumer receivers. The implementation will most likely remain restricted to special alert radios to be used under specific well-defined conditions applying to closed user groups only.

2.3.8 Option for Adding an AM Radio Data System

Section 5 of the new RBDS standard serves as a "placeholder" for a future AM data system (AMDS). The development of an AM data system is still supported by the NRSC and has been studied for more than five years by the RBDS subcommittee without an agreement being found as to the modulation system to be used. The requirement, difficult to meet, is that any such system must be compatible with the C-Quam AM stereo system.

In Europe, tests in the 1980s showed that low rate data could be carried by low frequency (LF) and medium frequency (MF) transmissions. For many years, the BBC 198 kHz transmitter has used phase modulation of the carrier to achieve 50 bps, but tests on "Hochschule für Verkehrswesen Dresden" showed that it was possible to add a data stream of 200 bps using a phase modulation of ±15 degrees to an AM transmitter, with minimal degradation of received audio quality.

Since 1986, extensive field trials have been conducted to find an appropriate baseband coding to match the bit error conditions of AM channels in LF, MF, and high frequency (HF).

AMDS [7], as it has become known, owes its origin to these trials. During 1994, the EBU ad hoc group R1/AMDS prepared a comprehensive proposal for the standardisation of the baseband coding for AMDS, leaving the modulation system still undecided since the phase modulation that was used in the tests is incompatible with the C-Quam AM stereo system still used in the United States.

2.3.9 Location/Navigation Information Deleted

The 1993 version of the RBDS standard had defined type 3A groups for location/navigation information. However, since this specification left a large number of options open, in which studies were continued within the RBDS subcommittee, nobody had implemented this feature, which was of no interest in Europe.

The new RBDS standard deleted that kind of specification and redefined the type 3A group to the Open Data Application's IDentification (AID) as required in the new RDS standard. As a result, the type 3A group is completely identical in both RDS and RBDS.

2.4 Use of the RDS Logos

In the United States, the RDS logos (shown in Figure 2.4) are now a registered trademark of the National Association of Broadcasters (NAB). The reason for this registration is solely to protect the logos against other registrations by individuals seeking to exploit them for commercial advantage. The copyright of the logos is jointly owned by the BBC and the EBU. Initially, these logos were developed by the BBC, and in August 1988, they were first published by the EBU in their RDS newsletters.

Before the logo can be used in the United States on any product, manufacturers will need to perform a self-certification test on their equipment, and submit a self-certification application to the Consumer Electronic

Figure 2.4 The RDS logos. (*Source:* EBU.)

Manufacturers Association (CEMA). The cost is $100.00 per year, per manufacturer. Only equipment displaying the logo will require registration. Figure 2.5 illustrates an RDS product demonstration.

Note that consumer receivers must use the term "RDS" rather than "RBDS." The use of the RBDS term is only permitted when referring to the North American standard.

In Europe, there is no self-certification test required, and the RDS logos can be used freely on equipment that implements *any* of the RDS features.

2.5 Conclusions

RDS and RBDS both allow broadcasters and receiver manufacturers to decide which of the possible standardised features they wish to implement. It is in that context that the question arises as to whether RDS manufacturers can adapt their software to modify existing products that were designed for Europe to create products for sale in the United States.

The answer to this question is not very obvious because of the large range of implementation options that are theoretically permitted in the two standards. However, in reality, the problem is not too difficult to be solved simply because most existing RDS car radios have implemented only the five basic features already mentioned above, and these also appear to be quite sufficient during the initial phase of the RBDS implementation in the United States.

Figure 2.5 RDS product demonstration in the United States. Note: RBDS is the name of the standard, but products shall be marked with the RDS logo. (*Source:* EBU.)

RDS and RBDS have in common exactly the same data broadcast signal modulation characteristics. This is the suppressed 57 kHz subcarrier that is PSK modulated with a data stream of 1187.5 bps and has identical baseband coding—the data synchronisation being optimised for mobile reception. Consequently, the same hardware for data demodulation and display can be used, and suitable radio data decoder ICs are currently available from Philips Semiconductors, SGS-Thomson, and others. Prices for these ICs are very competitive because several millions of these chips are now being made each year. Consequently, the question of adaptation becomes entirely a matter of software.

Depending on the feature set of a particular receiver, some differences may be discounted. However, "Caution should be exercised when making such decisions," said the chairman of the RBDS subcommittee, Scott Wright (Delco Electronics Corp.), when he analysed the differences of the new RDS/RBDS standards [8,9].

The essential differences between RDS and RBDS that have an impact on the design of consumer receivers are listed below. Recognition of these differences is essential to all manufacturers interested in a global market for RDS equipment and consumer receivers. The main differences are as follows:

- *Programme TYpe definitions (PTY):* The RBDS standard calls out a different set of programme type terms.

- *Programme Identification coding (PI):* North American PI codes differ in functionality: This affects alternate frequency switching and regionalisation.

- *Dynamic Programme Service (PS) name:* The RBDS standard differs in its allowance of "nondistracting" changes to the PS.

- *Phase out of fast Programme Service (PS) feature:* Type 15A group of RBDS was previously defined for use as a fast PS feature. This usage is being discontinued.

- *Multiplexed MMBS:* This is used as a commercial nationwide paging system in the United States. Since it shares the same modulation format and data protocol as RDS, it may be time division multiplexed with RDS. MMBS uses the offset word E (see also Appendix B), which is defined in both the RDS and RBDS standards to permit this multiplexing to take place without disturbing synchronisation of the received RDS/RBDS data stream.

If a receiver designer aimed at adapting existing RDS receiver designs for use in North America and made a receiver comply with the RBDS standard, all the above, as we have seen, indicates that the software changes that are needed are relatively small. Recognition of the offset word E (all zeros) would definitely be the most important change to be made, and when PTY is implemented in the receiver, the RBDS code table has to be used, of course. Finally, the PI codes have to be interpreted differently as far as AF switching and regionalisation are concerned (see Appendix F). That is practically all that is needed to make RDS receivers work correctly in North America (i.e., United States, Canada and Mexico).

References

[1] CENELEC, Specifications of the radio data system; EN 50067:1992, European Committee for Electrotechnical Standardisation, Brussels, Belgium, 1992.

[2] EIA/NAB National Radio Systems Committee, United States RBDS Standard, Specification of the Radio Broadcast Data System (RBDS), Washington, D.C., Jan. 8, 1993.

[3] CENELEC, Specification of the radio data system (RDS) for VHF/FM sound broadcasting, EN50067:1998, European Committee for Electrotechnical Standardisation, Brussels, Belgium.

[4] EIA/NAB National Radio Systems Committee RBDS Subcommittee-United States RBDS Standard, Draft 2.0, Specification of the Radio Broadcast Data System (RBDS), Washington, D.C., Aug. 1997.

[5] RDS Forum, Document 96/007: Comments on Programme Service name (PS)-Nonstandard transmission of type 0 group. European Broadcasting Union/RDS Forum, 17A, Ancienne Route, CH-1218 Geneva, Switzerland, 1996.

[6] Swedish Telecommunication Administration (Televerket)-Paging Receiver for the Swedish Public Radio Paging System, 76-1650-ZE, 1976.

[7] EBU, Proposal for a Data System for AM transmitters, EBU Document GT R1/AMDS 009, GT R1 910, European Broadcasting Union, 17A, Ancienne Route, CH-1218, Geneva, Switzerland, Sept. 9, 1994.

[8] Wright, S., *The Broadcaster's Guide to RDS*, Focal Press, 1997.

[9] EIA/NAB, Wright, S., "RBDS versus RDS: What are the differences and how can receivers cope with both systems?" 1997.

[10] Beale, T., D. Kopitz, "RDS in Europe, RBDS in the USA - What are the differences and how can receivers cope with both systems?" *EBU Technical Review*, No. 255, 1993.

3

RDS Features Serving as Tuning Aids

3.1 Introduction

In this chapter, RDS features that help the tuning functionality of FM receivers are introduced. In this connotation, a feature in RDS is considered to be something that gives a distinctive functionality. Originally in the earliest versions of the specification, these tended to be defined into a single RDS group such as RadioText, which is described in Chapter 4 (although there are notable exceptions such as PI, which is described in this chapter). But, as the specification was enhanced, some features were defined that make use of more than one group to be transmitted (e.g. Radio Paging, which is described in Chapter 8) to obtain full functionality.

Many of these features are considered to be included in the minimum set of RDS features that would be implemented initially by a transmission operator.

3.2 Basic RDS Features

The RDS specification, EN 50067:1998, defines a number of RDS group types that are generally reserved for a particular application. The type 0A group—basic tuning and switching information—is a special case, designed to carry the fundamental components of RDS all together in a single group that will be transmitted frequently to convey many pieces of information to an RDS receiver to enable it to perform a considerable number of tuning functions. These fundamental components, known as features, allow a degree of automatic tuning in an RDS receiver and allow it to present tuning-related information to the user. The type 0A group contains all of the following RDS features: Alternative Frequency (AF), Decoder Information (DI), Music Speech (MS), PI, Programme Type (PTY), Traffic Announcement (TA), and Traffic programme (TP). The type 0B group contains the same features except the AF feature [1].

The implementation of the type 0 groups in any RDS transmission is essential, and not optional—unlike many of the other possible groups in RDS—because it carries vital information required for automated tuning of an RDS receiver.

From these basic RDS features serving as tuning aids, it is useful to initially consider AF, PI, PS, and TP. They can be seen in various locations within the type 0A group, as shown in Figure 3.1. Furthermore, PI and TP can be found in *all* other groups to provide immediate information to an RDS receiver from every single group decoded.

The type 0B group has a very similar structure, but there is a simple difference in block 3, as shown in Figure 3.2. In block 3, the AF feature is not used, so it is replaced by a second PI code. This group is normally used by single-service RDS transmissions, where there is no need for an AF list because there are, indeed, no alternates.

3.3 Programme Identification—PI

The PI code is transmitted in block 1 of *every* single RDS group. PI is a machine-readable code that is not intended for direct display. It is uniquely allocated to each individual programme service within a geographical area to enable it to be distinguished from all other programme services. (For more detail about PI coding, see Appendix F.) Irrespective of frequencies used, transmissions with identical PI codes in any given area must have the *identical* audio programme. A receiver may therefore freely switch between two different transmissions with the same PI code in order to track optimum reception, assured that programme audio will be identical on both transmitters.

One important application of the information contained in the PI code is to enable an RDS receiver to automatically search for an alternative frequency in case of bad reception of the programme to which the receiver is tuned; the criteria for the changeover to the new frequency would be the presence of a better signal having the *same* PI code.

The PI code consists of 16 bits and is usually referred to by four hexadecimal characters, or nibbles, as shown in Figure 3.3.

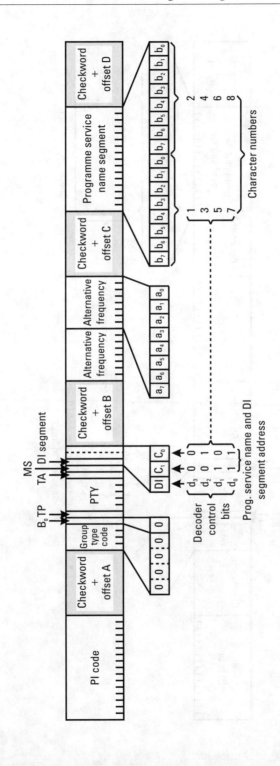

Figure 3.1 Type 0A group, showing the disposition of features carried. (*Source:* EBU.)

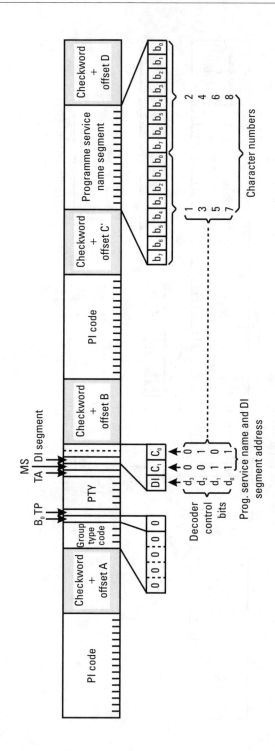

Figure 3.2 Type 0B group, showing the second PI code in place of the AF feature. (*Source:* EBU.)

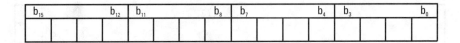

| b₁₅ | | | b₁₂ | b₁₁ | | | b₈ | b₇ | | | b₄ | b₃ | | | b₀ |

Figure 3.3 PI code structure. (*Source:* EBU.)

PI codes assigned to programme services are invariant, and other than changing the second nibble (regional element) in accordance with the specific rules, it must not be changed as part of regular radio programming. Furthermore, when cross-referring to other services, the PI code of referenced services is transmitted in block 4 of type 14 groups (see Chapter 5).

3.3.1 Broadcasting Conventions

PI codes must be allocated so that two different services do *not* transmit the same PI code in the same area. As part of this allocation process, the first nibble (hexadecimal character) is a Country Identifier, signifying the country of origin of the broadcast (not necessarily the country in which a transmitter is located), as shown in Figure 3.4. The country/area code allocations are given in the RDS specification EN 50067:1998, Annexes D and N [1], and in Appendix G of this book. The PI code allocations for Europe are shown in Figure 3.5.

In Europe, most broadcast networks are in fact services comprised of more than one transmitter which, at times, separate or split in order to offer regional information or programmes. Each of these regions must be provided

Area coverage code	L	I	N	S	R1	R2	R3	R4	R5	R6	R7	R8	R9	R10	R11	R12
HEX	0	1	2	3	4	5	6	7	8	9	A	B	C	D	E	F

I = International — The same programme is also transmitted in other countries
N = National — The same programme is transmitted throughout the country
S = Supra-regional — The same programme is transmitted throughout a large part of the country
R1 to R12 = Regional — The programme is available only in one location or region over one or more frequencies, and there exists no definition of its frontiers
L = Local — Local programme transmitted via a single transmitter (only) during the whole transmission time

Figure 3.4 Coverage area: hex codes used in the second nibble of the PI code. (*Source:* EBU.)

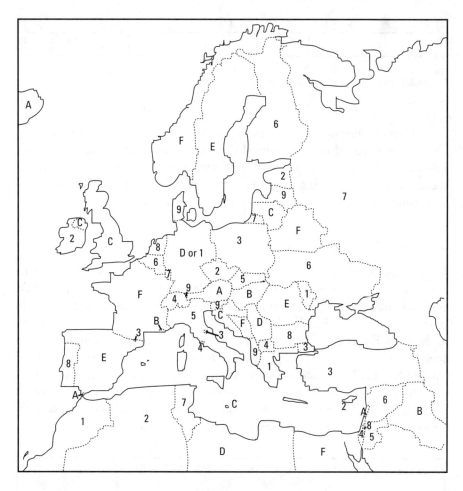

Figure 3.5 PI country code allocations for Europe. (*Source:* EBU.)

with a unique PI code by allocation of a different regional code in the second element. The regional element must be in the range 4-F hex. Services having identical first, third, and fourth elements, and a regional element in the range 4-F hex are said to be generically related or regional equivalent services. Up to 12 services, each using one of the regional codes (4-F hex) may be associated in this way. During the times when *all* regions are carrying common programmes, it is permitted for the services to change the regional element of the PI code to either 2 or 3 hex on each regional service. If the code is changed in this manner, all regions should ideally join/split at the *same* time. Partial splits/joins will cause undesirable effects for some listeners. This is the only change of PI code normally permitted. Changing PI codes is *not* permitted on programme services used as part of an Enhanced Other Network (EON).

Although within a country code area, PI codes *may* be reused at geographically separated areas, it is preferable that each service be allocated a unique code. RDS receivers use the PI code as addresses against which to store programme attributes (Programme Service name, list of alternative frequencies, etc.). Use of the same PI code by more than one programme service would probably cause data about one service to be misassociated with another having the same PI code.

To prevent having to reuse PI codes and yet still be able to associate together more than 12 regional equivalent services, the Extended Generic (EG) feature should be used. The word generic has a somewhat special meaning in RDS and involves some very complex ideas, which require a very full appreciation of how broadcasters view their services. Furthermore, sometimes they transmit common programmes in a network and this must be signalled to an RDS receiver. A more detailed explanation can be found in the EBU/RDS Forum document BPN 009: RDS Guidelines [2].

In this case, PI codes that have the extended generic indicator (EG) flag set, and are identical in the first and third elements only, and regional elements in the range 4-F hex, are also regional equivalent services to each other. Up to 192 (12 × 16) such services may be grouped in this way.

3.3.2 Reception

The PI code is arguably the single most important feature of RDS. It uniquely identifies a programme service, within any geographical area, where broadcasts sharing the same PI code are guaranteed to be carrying identical programme audio. Although there are many designs possible for an RDS receiver, evaluation of the PI code is fundamental to operation.

In any receiver with preset memories, it is essential for the PI code of the broadcast to be stored in nonvolatile memory when a service is assigned to a memory location. If no signal with the correct PI code is available, on the last tuned frequency or AFs, when a preset is chosen, the receiver should scan the FM band, stopping on each receivable RDS service and evaluating the PI code.

The correct transmission to select is that which gives the best signal with a PI code matching that recalled from memory. If no service with a matching PI code can be found, a receiver should select a transmission with (in order of priority) a PI code identical in the first, third, and fourth elements or a PI code identical in the first and third elements only. Although accepting these transmissions as regional variations of the wanted service, under no circumstances should the PI code stored in preset memory be overwritten by any other PI code.

An RDS receiver must always evaluate the PI code and confirm it is correct before it remains tuned to a different transmission. Usually, switching between two different transmissions should occur only when PI codes are identical in all four nibbles. Optionally, a receiver may provide a mode that enables receivers to switch automatically to other regional variants of the same service.

3.4 Programme Service (PS) Name

The PS name is an eight-character alphanumeric label transmitted by the broadcaster to identify to the listener the name of the tuned service. It is not intended for automatic tuning purposes. The PS name is transmitted two characters at a time in either the type 0A or type 0B groups. (See Figures 3.1 and 3.2 above.) A complete PS name requires a minimum of four groups, which at the recommended rate takes just one second to be transmitted. As the reception of a PS name takes a minimum of one second, and even longer in adverse reception conditions, receivers may store the PS name to allow instant recall upon the user accessing a preset memory. Since broadcasters may choose to transmit PS names in either type 0A or 0B groups, receivers must be capable of using both transmission methods.

PS names can also be carried as part of the EON data and then provide the PS name of the cross-referenced programme services.

The PS name should only be used as a static feature, but it may occasionally change. For example, regional services may carry their own PS name during the day when they originate their own programmes. However, the PS name is changed when they are transmitting a common programme originated by one of these broadcast services during the evenings and nights.

Due to these design parameters for PS, the use of the PS feature to transmit text other than a single eight-character representation of the Programme Service name is not permitted. If a broadcaster wishes to transmit more programme service information, programme-related information, or any other text, then the RadioText feature provides the capability to achieve dynamic text possibilities. The RDS Forum has been instrumental in keeping media regulators informed about the correct use of PS, which has, unfortunately, been incorrectly implemented by several broadcasters who have not realised the importance of these design parameters for PS [3].

Appropriate use of spaces, punctuation, and combinations of uppercase/lowercase characters and accented characters should all be used to improve readability of the PS name. For example, in the Untied Kingdom, the BBC uses >_BBC_ R4_< for the BBC Radio 4 network PS and classic FM uses >Classic_< for its PS. The characters to be used for PS are shown in the RDS specification EN 50067:1998 Annex E, which indicates a "limited character set" that may only be available to a low-cost receiver, using only uppercase characters together with numbers and a few punctuation marks. The limited character set for display is indicated in a footnote to Figure E1 in Annex E of the RDS specification (see Appendix I, Figure I.1). Receivers capable of only displaying the limited character set must convert the lowercase characters in columns/rows 6/1–7/10 to their column 4 and 5 uppercase character equivalents and convert accented characters in columns 8-15 to their non accented equivalent. For example, a low-cost receiver would need to convert the received character "a" to "A" for display; it would also have to convert "å" to "A" and convert other characters such as "é" to "E." A midrange receiver displaying lowercase characters in columns 6 and 7 may still be required to convert received accented characters such as "ñ" to "n" for display. Clearly, receivers capable of displaying the full character set will provide a superior presentation.

Broadcasters have to be aware that not all receivers' displays will support the full code table, and may only support the limited character set.

Even though PS has constrained design parameters, to achieve very reliable PS reception and to provide a safe invariant display, it can be employed in a receiver to offer exceptionally user-friendly performance. For example, a dual front-end tuner car radio from Becker (see Figure 3.6) uses incoming PS information to derive and generate a three-character soft label for a number of preset buttons. This allows the user to see a description of other radio stations with some identification, which is better than seeing a preset number label such as "1," "2," "3," and so on.

Figure 3.6 Becker car radio using PS-derived soft labels for programme selection buttons. (*Source:* Becker.)

3.5 Alternative Frequency List—AF

The purpose of the AF list is to facilitate the automatic tuning of an RDS receiver. The AF list indicates carrier frequencies on which the specified service is broadcast. AF lists may be broadcast for the tuned service and also for other services using the EON feature. Generally, at any one location, only a few (or even none) of the frequencies indicated will carry the intended service at good signal strength. Receivers must ascertain the suitability of each frequency in the list by evaluating the signal quality and then decoding the PI code to check that there is an exact match before switching frequencies automatically.

AFs for the tuned network are broadcast two codes at a time, in block 3 of type 0A groups, according to codes given in Tables 3.1 and 3.2.

AF lists for cross-referenced services are carried in variants 4, 5, 6, 7, 8, and 9 of type 14A groups.

Ideally, AF lists should be as short as possible and contain *only* the frequencies of immediately adjacent transmitters. Broadcasters may achieve this by installing RDS encoders at each transmitter site, with an individual appropriate list of frequencies. This coding method, known as *Method-A*, is the most widely used and suitable for lists not exceeding 25 frequencies. At the recommended repetition rate for the transmission of type 0A groups (i.e., four groups per second), a list with the maximum of 25 frequencies will take less than four seconds to transmit in its entirety. Figure 3.7 shows an example Method-A list and explains the format when an even number of frequencies are in the list, necessitating the use of the filler code (which can be ignored by the receiver).

A second way to transmit AFs is provided where it is not possible to restrict the list to fewer than 25 frequencies, or where it is required to indicate frequencies that belong to different regions that at times carry different programmes. In this coding of alternative frequencies, known as *Method-B*, all

Table 3.1
VHF Code Table

Number	Binary Code	Carrier Frequency
0	0000 0000	Not to be used
1	0000 0001	87.6 MHz
2	0000 0010	87.7 MHz
:	:	:
:	:	:
204	1100 1100	107.9 MHz

Table 3.2
Special Meanings Code Table

Number	Binary Code	Special Meaning
0	0000 0000	Not to be used
205	1100 1101	Filler code
206	1100 1110	Not assigned
:	:	:
223	1101 1111	Not assigned
224	1110 0000	No AF exists
225	1110 0001	1 AF follows
:	:	:
249	1111 1001	25 AFs follow
250	1111 1010	An LF/MF frequency follows
251	1111 1011	Not assigned
:	:	:
255	1111 1111	Not assigned

transmitters broadcast in sequence, a series of transmitter-related lists that are organised in frequency pairs, with the particular transmitter frequency always repeated in each pair as a reference together with one of the respective alternative frequencies of that transmitter.

The particular way in which the data is organised within block 3 of type 0A groups enables a receiver to deduce whether Method-A or Method-B coding is being used and, in the case of Method-B, to derive the shorter lists

F_1	F_2	Explanations
# 6	89.3	Total number (6) of frequencies in this list (excluding the Filler code) 89.3 MHz is the frequency of a "main transmitter"
88.8	89.0	88.8 MHz and 89.0 MHz are AFs of tuned frequency (possibly re-broadcast transmitters)
99.5	101.7	99.5 MHz and 101.7 MHz are AFs of tuned frequency (possibly re-broadcast transmitters)
102.6	"F"	102.6 MHz is an AF of the tuned frequency (possibly re-broadcast transmitter) and Filler code

Figure 3.7 Example of an AF list using Method-A. (*Source:* EBU.)

appropriate to each transmitter within the network. As the complete list of frequencies carried by Method-B may be very long, a complete cycle of information may take up to two minutes (or even longer in poor reception conditions) for a receiver to acquire.

Clearly, receivers must be able to detect and use AF lists coded in either the Method-A or Method-B protocols. The rules governing Method-B transmissions are complex and beyond the scope of this book, but Figure 3.8 gives a flavour of the possibilities.

The presence of a frequency within an AF list does not guarantee that the frequency will provide a suitable signal in any particular area, and especially does not guarantee that a transmission on a frequency will be that intended by the broadcaster. In areas where frequency planning or regulation of broadcasting is inadequate, a particular frequency may be reused for quite different services in close proximity. It is essential therefore that the receiver, before accepting a new frequency, tunes to the intended frequency to check and confirm the PI code is the one expected. Only if the PI code matches the one intended, should the receiver accept the frequency as the new tuned frequency. Receivers should reject frequencies where there is no detected PI code or the wrong PI code.

When retuning to an alternative frequency, a receiver will inevitably produce, in the case of a single front-end unit, a short audio interruption on the

F_1	F_2	Explanations
# 11	89.3	Total number (11) of frequencies for tuning frequency: 89.3 MHz
89.3	99.5	$F_2 > F_1$: 99.5 MHz is an AF of tuned frequency 89.3 MHz and is the same programme
89.3	101.7	$F_2 > F_1$: 101.7 MHz is an AF of tuned frequency 89.3 MHz and is the same programme
88.8	89.3	$F_2 > F_1$: 88.8 MHz is an AF of tuned frequency 89.3 MHz and is the same programme
102.6	89.3	$F_2 < F_1$: 102.6 MHz is an AF of a regional variant of tuned frequency 89.3 MHz
89.3	89.0	$F_2 < F_1$: 89.0 MHz is an AF of a regional variant of tuned frequency 89.3 MHz
# 9	99.5	Total number (9) of frequencies for tuning frequency: 99.5 MHz
89.3	99.5	$F_2 > F_1$: 89.3 MHz is an AF of tuned frequency 99.5 MHz and is the same programme
99.5	100.9	$F_2 > F_1$: 100.9 MHz is an AF of tuned frequency 99.5 MHz and is the same programme
104.8	99.5	$F_2 < F_1$: 104.8 MHz is an AF of a regional variant of tuned frequency 99.5 MHz
99.5	89.1	$F_2 < F_1$: 89.1 MHz is an AF of a regional variant of tuned frequency 99.5MHz

Figure 3.8 Examples of AF lists using Method-B, along with their meaning. (*Source:* EBU.)

order of a few milliseconds. Checking a PI code, however, will take at least 100 milliseconds, during which time it is possible for a burst of incorrect programme to be heard, should the particular frequency being tested be other than the intended programme. Manufacturers should adopt an intelligent process as a background activity to validate each frequency in the AF list. This will ensure that a receiver has available for use, when required, a frequency with an optimum audio quality and with the correct PI code. Unwanted bursts or breakthrough from other programmes (of the wrong PI code) and audible mutes should be avoided. Dual-tuner receivers may have an advantage in AF list evaluation as the checking of PI codes may be done by the second tuner while the first tuner continues to provide an uninterrupted audio programme for the listener.

Some broadcasters may include frequencies of services in their AF lists—simultaneously broadcast on AM perhaps, where there is no suitable FM transmission. In this case, it is not possible to confirm the frequency that carries the intended programme as no PI code is transmitted on normal AM frequencies. Use of AM frequencies is probably only viable where the AM tuner part of the receiver is separate from the FM one. This would allow a receiver, while the AM tuner is providing an audio programme, to use the FM tuner to perform RDS processing and remain active to allow the earliest possible return to an FM transmission.

3.6 Traffic Programme (TP) Flag

The RDS traffic information service allows listeners to receive traffic reports even if listening to a CD/cassette, or with the receiver volume turned down or muted. This service makes use of two flags: the TP and TA flags. The latter is described more fully in Chapter 6; here, the tuning aspect of the TP flag is considered. The TP flag is carried in block 2 of *every* group type and in block 4 of type 15B groups. If the TP flag is set to /1/, this indicates that the tuned programme service provides the RDS traffic service. Broadcasters must *not* set the TP flag to /1/ unless the TA is also *dynamically* controlled [1].

Programme services that do *not* provide traffic services with a switched TA flag, but instead cross-reference via EON services that do, indicate this by setting the TP to /0/ and the TA to /1/ on the service. When a listener selects the RDS traffic service feature, the receiver uses the status of both the TP and the TA flags to determine whether or not the tuned service can provide the traffic information service itself, or alternatively in the case of an EON-capable receiver, via a cross-referenced service.

The TP flag can be used by an RDS receiver, since it is in every RDS group, to easily evaluate the availability of the RDS traffic service when checking a frequency as part of the automatic tuning capability. If the tuned service does not provide the RDS traffic service feature, an RDS receiver should indicate this to the listener in some way. Options include the following:

- Visually on the display or audibly by a bleep. This is particularly useful when going away from a service and signal strength weakens to the point where the TP flag status is becoming uncertain.

- The receiver may automatically start to seek for an alternative service able to provide a traffic service.

Alternatively, the receiver could allow the user the option of continuing to listen to the currently tuned service without the traffic service, or manually initiating a search for an alternative service. The options provided might be different, depending on how the user is currently operating the receiver. For example, a routine may be as follows:

- If the RDS traffic information service is selected while the user is listening to a CD/cassette, the receiver may automatically seek and tune to a service capable of providing a traffic service, without either beeping or requiring a listener to manually initiate a search.

- If listening to the radio when traffic information is selected, the receiver should alert the listener visually or with a brief audible warning if the traffic service is unavailable, but not automatically retune. The user then has the option to initiate a seek for another traffic service programme, if required, or to continue listening to the tuned service but without a traffic service.

The same routine should be adopted if, while tuned to a service that was offering a traffic information service, the status of the TP/TA flags changes and indicates that a travel service is no longer available. Interference on areas of poor reception will mean that the TP and/or the TA flag may at times be read unreliably. It is important that the status of TP and TA flags be evaluated over several groups, and that momentary switching be avoided as the result of short-term signal fluctuations.

A full description of the interaction of TP flags, TA flags, and EON is given in Chapter 6.

3.7 Slow Labelling Codes

The RDS specification contains some additional slow labelling codes that are used to support various features. These are specified to be carried in the type 1A group at a much slower rate than the main feature groups; for example, once per minute. The type 1A group also contains the functionality of the PIN feature (see Chapter 5), as shown in Figure 3.9.

3.7.1 Extended Country Code (ECC)

The ECC feature was built into the RDS specification some time after the original publication of the specification, because originally it had been designed only for Europe and the limited number of country identifier codes (only 15 different codes, 1-F hex) could easily be shared across the geographical area considered. Note, for example, in Figure 3.5 that E hex is shared between Sweden and Spain. The use of RDS has grown considerably since then and, indeed, the need to accommodate newer emerging countries has resulted in the ECC feature, which is now an important part of the full implementation of several other features. When widely implemented, this feature allows receivers to initialise automatically for country-specific applications; therefore, broadcasters are being encouraged to implement this feature.

The ECC feature, when read in conjunction with the country code part of a PI code, forms a *unique* Country Identifier. This may be used by a receiver to automatically select country-specific applications; for example, PTY language tables, which are different according to the broadcaster's choice of RDS or RBDS. To support this functionality, the ECC, which consists of 8 bits transmitted in block 3 of type 1A group's variant 0 (as shown in Figures 3.9 and 3.10), should be transmitted at least once per minute.

Returning to the example of Spain and Sweden, Romania also uses E hex in the allocated country PI code. However, each country is assigned a different Extended Country Code: Romania = E1, Spain = E2, Sweden = E3, which taken together with the country code element of the PI code provides a unique Country Identifier. Full listings of the ECC and PI country/area codes are given for European countries in EN 50067:1998 Annex D, and for the rest of the world in Annex N (see also Appendix G of this book) [1].

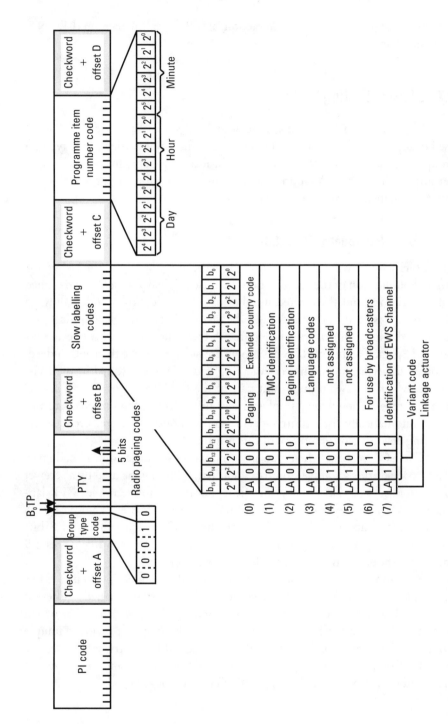

Figure 3.9 Type 1A group showing among other slow labeling codes the ECC and language identification features. (*Source:* EBU.)

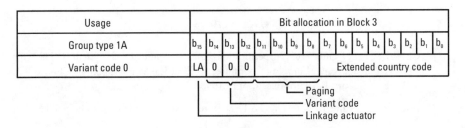

Usage	Bit allocation in Block 3															
Group type 1A	b_{15}	b_{14}	b_{13}	b_{12}	b_{11}	b_{10}	b_9	b_8	b_7	b_6	b_5	b_4	b_3	b_2	b_1	b_0
Variant code 0	LA	0	0	0					Extended country code							

Paging — Variant code — Linkage actuator

Figure 3.10 ECC bits in type 1A groups, variant 0. (*Source:* EBU.)

3.7.2 Language Code

Another enhancement that remains optional, but which is easy to transmit and could lead to enhanced receiver performance, is the language code, also carried using 8 bits in type 1A groups, variant 3 (as shown in Figure 3.11). Language identification may be used to signal the spoken language of the current programme on a service. The feature may be of interest not only in multilingual countries but also for visitors to help locate programmes in their preferred language.

Broadcasters may automatically control the language code by including a data field in their RDS programme scheduling system, and this could then be used to control their RDS encoders automatically. They should be aware that the normal repetition rate for type 1 groups to convey the PIN feature is approximately once per minute and that the expected rate for language identification will go down if several variants of type 1 groups are utilised. It is, however, recommended to transmit variant 3 if the language code is implemented at least once every two seconds. This makes the feature unsuitable for search tuning.

The language codes, described as two hex numbers, are specified in EN 50067:1998, Annex J [1].

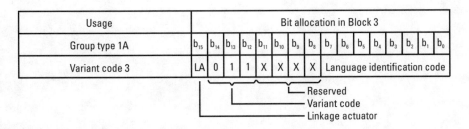

Usage	Bit allocation in Block 3															
Group type 1A	b_{15}	b_{14}	b_{13}	b_{12}	b_{11}	b_{10}	b_9	b_8	b_7	b_6	b_5	b_4	b_3	b_2	b_1	b_0
Variant code 3	LA	0	1	1	X	X	X	X	Language identification code							

Reserved — Variant code — Linkage actuator

Figure 3.11 Language code bits in type 1A group, variant 3. (*Source:* EBU.)

References

[1]   CENELEC, EN 50067:1998-Specifications of the Radio Data System (RDS.) for VHF/FM broadcasting, European Committee for Electrical Standardisation (CENELEC), 35B rue de Stassart, B-1050 Brussels, April 1998.

[2]   EBU /RDS Forum: BPN 009-Radio Data System (RDS) Guidelines, to be published in Autumn 1998.

[3]   RDS Forum Statement, sent with letter DT/969-S/DK dated Feb. 26, 1996, RDS Forum Statement, sent with letter DT/3255-S/DK dated Dec. 9, 1997

4

Radio Programme-Related RDS Features

4.1 Introduction

This chapter provides a description of the original features and some new features that are related to the programme content in the audio channel of the FM transmission.

RDS was originally designed with a strong view that it should provide features that would let the user benefit from information related to broadcast programmes. In Chapter 3, it was noted that type 0A groups contain both tuning features and programme-related features. One of these programme-related features, Programme TYpe (PTY), was foreseen by the original developers to be significant for tuning purposes, so PTY information is conveyed in *all* RDS groups. But PTY did not become an early success for RDS; nevertheless, it is being implemented by more broadcasters and receiver manufacturers alike now that the capabilities offered are better understood—and now that other features are also demanding dynamic control over the RDS encoders. More recently, the possibilities of an additional feature, PTYN, has been developed and added to the RDS specification. [1].

PTY can be used to help the user with the programme selection process and so too can the PIN feature. Other dynamic features to be discussed in this chapter are RadioText—which has many uses, giving both programme-related information and many other possibilities. Finally, the support features Decoder Identification/Programme TYpe Identification (DI/PTYI) are described.

4.2 Programme TYpe (PTY)

The PTY codes are transmitted, using five bits in block 2 of *every* single RDS group. Additionally, the type 15B group—providing fast basic tuning and switching information—carries a second opportunity for decoding the PTY code, which is also transmitted in block 4 (see Figure 4.1). Furthermore, the code may also be carried in type 14A groups, variant 13 from transmissions cross-referencing a programme service.

The PTY codes are given in the RDS specification EN 50067:1998, Annex F [1]. These codes provide standardised descriptions for broadcasters to depict their programmes according to a short list of 29 standardised choices. Although simple in nature, this feature has been complex for broadcasters to implement because they do not find it easy to narrow their programme descriptions down to a short list. Indeed, it is claimed to be a living issue, with categories gradually changing with broadcasting trends and styles. Table 4.1 shows the PTY codes, decimal number, the five bits listed, and a very short description for each, together with a recommendation for both 8- or 16-character displays in English. The RDS specification has introduced the possibility of 16-character displays now that receiver technology is more able to use so many characters. Displays in other languages have been developed and agreed upon by experts for many languages through the RDS Forum, and are detailed in Appendix H. In Chapter 2, Table 2.2 shows the RBDS PTY codes list adjacent to the RDS codes list—the differences in the meanings of the same transmitted codes must be allowed for by RDS receivers being built for the RBDS or RDS markets, or by a receiver making use of automated code table switching (when it can detect the ECC feature, which is described in Chapter 3).

Great care is needed when considering PTY coding in order to conform with either one specification or the other (e.g., RBDS or RDS). This is because the same defining words intended for display, such as Oldies Music, have subtly different meanings according to the specifications and different codes used (i.e., RDS uses PTY code 27 for Oldies and RBDS uses PTY code 11) [1, 2].

As shown in Table 4.2, three of the PTY codes are harmonised in both RDS and RBDS, since they have special meaning.

Although any PTY code can be used in any mode, some categories are more suited to one particular mode of operation than another. Programmes that are generally listened to in their entirety (to receive coherent information) are best matched to receivers offering the PTY-STANDBY mode of operation. This applies to many speech programmes such as news, weather, and drama.

Some broadcasters use PTY to indicate the general nature of their service as a whole, with no changes at programme junctions (static PTY); others assess

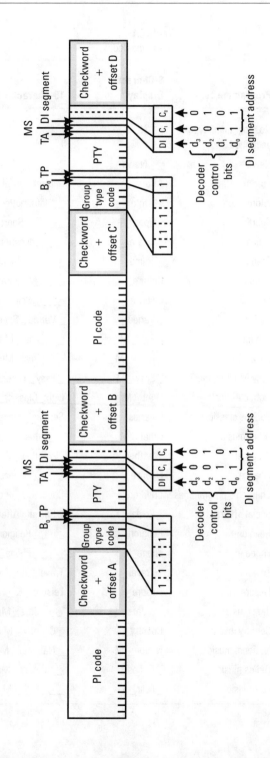

Figure 4.1 Type 15B group, showing the PTY bits. (*Source:* EBU.)

Table 4.1
Programme Type Codes

Number	Code	Programme Type	8-Character Display[1]	16-Character Display[1]
0	00000	No programme type or undefined	_ _ None _ _	_ _ _ _ _ _ None_ _ _ _ _ _
1	00001	News	_ _ News _ _	_ _ _ _ _ _ News_ _ _ _ _ _
2	00010	Current affairs	Affairs _	Current _ Affairs _
3	00011	Information	_ _ Info _ _	_ _ Information _ _ _
4	00100	Sport	_ Sport _ _	_ _ _ _ _ Sport _ _ _ _ _ _
5	00101	Education	Educate _	_ _ _ Education _ _ _ _
6	00110	Drama	_Drama _ _	_ _ _ _ _ Drama _ _ _ _ _ _
7	00111	Culture	Culture _	_ _ _ _ Cultures _ _ _ _
8	01000	Science	Science _	_ _ _ _ Science _ _ _ _
9	01001	Varied	_ Varied _	_ Varied _ Speech _ _
10	01010	Pop music	_ Pop _ M _ _	_ _ _ Pop _ Music _ _ _ _
11	01011	Rock music	_ Rock _ M _	_ _ _ Rock_Music _ _ _
12	01100	Easy listening music[2]	_ Easy _ M _	_ Easy _ Listening _
13	01101	Light classical	Light _ M _	Light_Classics_M
14	01110	Serious classical	Classics	Serious _ Classics
15	01111	Other music	Other _ M _	_ _ Other _ Music _ _ _
16	10000	Weather	Weather _	_Weather _ & _ Metr _
17	10001	Finance	Finance _	_ _ _ _ Finance _ _ _ _ _
18	10010	Children's programmes	Children	Children's _ Progs
19	10011	Social affairs	_ Social _	_ Social _ Affairs _
20	10100	Religion	Religion	_ _ _ _ Religion _ _ _ _
21	10101	Phone In	Phone_In	_ _ _ _ Phone _ In _ _ _ _
22	10110	Travel	_ Travel _	Travel _ & _ Touring
23	10111	Leisure	Leisure _	Leisure _ & _ Hobby _
24	11000	Jazz music	_ _ Jazz _ _	_ _ _ Jazz _ Music _ _ _
25	11001	Country music	Country _	_ Country _ Music _ _
26	11010	National music	Nation _ M	_ National _ Music _
27	11011	Oldies music	_ Oldies _	_ _ Oldies _ Music _ _
28	11100	Folk music	_ Folk _ M _	_ _ _ Folk _ Music _ _ _

Table 4.1 (continued)

Number	Code	Programme Type	8-Character Display[1]	16-Character Display[1]
29	11101	Documentary	Document	_ _ Documentary _ _ _
30	11110	Alarm test	_ _ TEST _ _	_ _ _ Alarm _ Test _ _ _
31	11111	Alarm	Alarm _ ! _	Alarm _ - _ Alarm _ ! _

1) These short terms are recommended for the 8- or 16-character display of the radio in English. Other language versions are available from the EBU and the RDS Forum.

2) In earlier versions of the RDS specification, the term MOR music was used for this code, but Easy listening is now more frequently used.

The symbol "_" is used to represent a space in the display

Table 4.2
Special PTY Codes and Meanings

Number	Programme Type	Description
0	No programme type or undefined	Indicates a particular programme is undefined
30	Alarm test	Indicates a test transmission and may be used by a consumer receiver for display
31	Alarm	Indicates a transmission of great importance, carrying information about a national or local emergency and must always be combined with a spoken message

PTY codes 1–29: Allow receivers to offer several modes of operation, which have been more clearly defined in the last few years.

PTY-SEARCH: Allows a listener to request the receiver to find a particular type of programme, if available.

PTY-SELECT: Allows a listener to request that the receiver indicate which particular Programme TYpes are currently available, and which could be selected.

PTY-STANDBY (or WATCH): Allows a listener to request that the receiver standby or watch out for a particular Programme TYpe to begin, switching to it when it does.

PTY-STORE: Allows a listener to request that the receiver (incorporating a means of recording audio) store a programme by matching requested PTY code, allowing later listening by the use.

and use PTY on a programme item by programme item operational basis (dynamic PTY).

The PTY-STANDBY mode of operation of a receiver is only viable with dynamic PTY operation, which must be flagged in the DI /PTYI feature (see Section 4.5).

When broadcasters use dynamic PTY, two codes (01—news and 16—weather) are principally expected to be used for short-duration programmes of no more than 5 or 10 minutes. Figure 4.2 shows a car receiver with a news selection button. They should not be used to describe longer programmes, such as 30-minute news bulletins that inevitably include much comment (and would be more appropriately coded as current affairs). Similarly, a longer feature programme about the weather should be coded as science.

Dynamic control of the PTY code requires that a broadcaster have a means for controlling the PTY code via a data connection to the RDS encoders at each transmitter. Ideally, PTY codes should be controlled directly from the on-air studio in real time, but control from a computer-based programme schedule is also possible provided audio programmes run completely and accurately on time.

Broadcasters operating more than one programme service that are cross-referenced via EON need to consider the interaction between services when controlling and allocating PTY codes. For example, at the "top of the hour," more than one service may transmit a news bulletin. The broadcaster should decide whether it is appropriate for each service to be signalled as news simultaneously or whether only one service at a time should be designated as the news service for the benefit of receivers in the PTY-STANDBY mode of operation. The use of PTY should always be made with the interests of the listener in mind.

Figure 4.2 Philips car receiver with PTY featuring a news selection button. (*Source:* Philips.)

4.2.1 PTY-SEARCH Mode

This is the mode most often provided on early RDS receivers, but on its own (i.e., without PTY-SELECTION or PTY-STANDBY modes, described below), it may have contributed to some customer dissatisfaction. In the PTY-SEARCH mode, a listener requests their desired PTY code by choosing from all the types specified and theoretically possible; the receiver then searches the band to see if the required PTY code is being broadcast at that time. When the requested code is unavailable, the listener may feel frustrated, and either has to request another choice or repeat the search until eventually successful.

4.2.2 PTY-SELECTION Mode

In this mode, a receiver presents the listener with a display of all PTY codes currently available on-air. On a receiver with a large display, all codes may be shown at once so that the listener is able to make an instant selection. On other receivers, the PTY codes available could be presented sequentially. The listener is able to make a selection from those codes actually available.

4.2.3 PTY-STANDBY Mode

This mode (also known as Watch or Wait mode) allows the listener to program a receiver to standby for a particular PTY code to become available. It may only be used reliably when the DI-d_3 bit is set to indicate that dynamic PTY is available. When tuned to a non-EON service (or using a non-EON receiver), the only possibility is for a receiver to monitor the PTY code on the "tuned" service and interrupt CD/cassette operation when the desired PTY code becomes available.

When tuned to an EON service (and using an EON receiver), the receiver should also monitor the PTY carried in block 3, variant 1 of type 14A groups being received, and this will allow the receiver to offer programmes matching the user's preselected PTY.

When the required PTY code is no longer available, the receiver may either return automatically to the original service or remain tuned to the current one. If the latter option is chosen, an easy means should be provided to allow a listener to return to the original service (e.g., pressing the PTY button). The same mechanism may be used to allow a listener to return to the original service partway through a programme (e.g., after the sports news headlines but before the sport in detail).

A receiver in standby mode should not switch across to a service once the programme has started (i.e., the receiver should only switch at a PTY transition).

4.2.4 PTY-STORE Mode

This is a further refinement of the PTY-STANDBY mode of operation, where programmes of the requested type are stored in memory or on tape, allowing a listener to replay them when requested. A dual-tuner RDS receiver could store the news bulletin (PTY code 01) on one service while the user is listening to a programme item on another service. The user then has the opportunity to listen to the news at a convenient time. A similar feature is possible with a single-tuner receiver while the listener has chosen to listen to a CD or cassette.

A receiver (dual-tuner) could offer a particularly attractive PTY implementation by combining all these modes of operation into a single routine. For example, say a listener requests a sports programme. The receiver *searches* for PTY code 04 (Sport). It is currently unavailable, so the receiver automatically goes into the PTY-STANDBY mode. The receiver then offers the listener the *selection* of categories currently available from which to choose while the receiver is standing by. The receiver switches across to a sports programme on a service when it begins. If no sports programme occurs while the user continues to listen, the receiver remains in the PTY-STANDBY mode and automatically *stores* sports bulletins when they occur, ready for the listener to replay on demand.

4.2.5 PTY-ALARM Function and Testing

PTY code 31, ALARM, indicates a transmission of great importance, carrying information concerning a national or local emergency. Therefore, all receivers should detect this code, sound an alarm, and turn up the volume.

Broadcasters need to occasionally test the broadcast chain to confirm that it can perform dynamic PTY switching and that they can fully use PTY code 30. This code is thought to be particularly useful in testing the system functionality without causing unnecessary public concern. Consumer receivers should neither use this code for searching nor to cause dynamic switching. Receivers may, if desired, display: >_TEST_< when detecting a PTY code 30 on a programme service.

4.3   Programme Type Name (PTYN)

The PTYN feature has been added to the RDS specification to allow a broadcaster to transmit an eight-character additional description that depicts the programme content. This can be used to enhance the PTY code description and is expected to help address the broadcaster's attitude that the standardised PTY codes are too limited.

It is intended to aid *selection* of listening more precisely than using the PTY code alone. It cannot be used in PTY-SEARCH, PTY-STANDBY, or PTY-STORE modes of operation (see Section 4.2).

The PTYN feature is transmitted in type 10A groups, four characters at a time; hence, two groups are required to complete the eight-character description. The use of PTYN to transmit text other than to a single eight-character description is not permitted, and it must not be used to provide sequential information. Figure 4.3 shows the type 10A group structure, which is designed to convey the eight characters for reconstruction into the normal eight-character display constraint of RDS receivers. The bit format follows that specified for RadioText, but only using eight characters (see Section 4.4).

The PTYN feature is capable of adding more precise descriptions about programme content than that provided by PTY alone and is particularly useful to distinguish between different services that use the same general PTY description. For example, if there are three services carrying PTY coded as SPORT, the PTYN feature can be used to add the descriptions: >Cricket_<, >_Golf_< and >_Tennis_< to each.

A single-tuner receiver could use PTYN to provide a second-level description of the tuned programme service. For example, after a listener makes a primary selection of SPORT, a receiver tunes to each of the services coded as SPORT in turn and is able to decode and present the PTYN description to the listener for final selection.

An attractive implementation in a dual-tuner receiver would be to use PTYN to provide a submenu. Such a receiver could present a complete list of both available PTY codes plus PTYNs, if any. For example, if there are three services broadcasting SPORT, the PTYN feature would provide details enabling a listener to chose his or her favourite sport directly from the menu.

4.4 RadioText (RT)

The RT feature is primarily designed for the broadcaster to transmit text messages of up to 64 characters at a time, for display by the fixed-location home receivers. As shown in Figure 4.4, RadioText uses blocks 3 and 4 in type 2A groups or block 4 in type 2B groups to transmit text messages of up to 64 and 32 characters, respectively. A further four bits in block 2 are used as address bits, and one bit is used as a Text A/Text B flag which toggles every time a new message is transmitted.

The message is transmitted four characters at a time in the type 2A groups and two characters at a time in the type 2B groups. The text is coded in

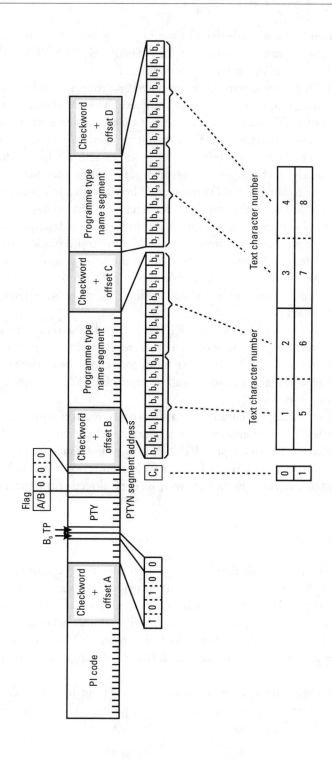

Figure 4.3 Type 10A group, showing the PTYN bits. (*Source:* EBU.)

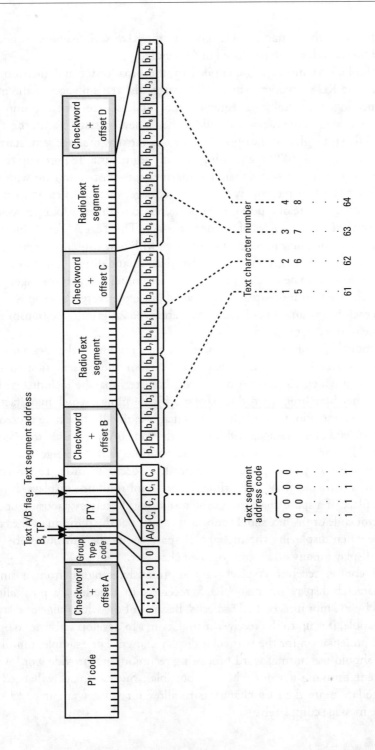

Figure 4.4 Type 2A groups showing the disposition of the RT feature. (*Source:* EBU.)

accordance with the character tables given in the RDS specification, Annex E. Some of theses codes are reproduced in Appendix I.

A RadioText message is compiled by the broadcaster and then downloaded to the RDS encoder, which will manage the transmission of the message, repeating it cyclically in sequence using at least one type 2 group per second for as long as the message is valid. When a new message is sent, the Text A/Text B flag is toggled to the opposite state and the new message sent, starting with address segment /0000/ and using as many segments as required up to 16. If the message requires fewer than 16, the message must terminate with the code 0D hex (carriage return). When a new message is sent, it is recommended that the repetition rate of type 2 groups be increased to three or four per second for at least the first two cycles of the new message. The Text A/Text B bit *must* remain static until a new message (i.e., message with new content) is sent.

To allow time for a receiver to receive, decode, and display a message, and to allow time for it to be read by the user, it is recommended that messages generally be changed no more frequently than about once every 10 seconds. This assumes each is transmitted at least twice—and type 2A groups are transmitted at the rate of three per second.

Although the primary responsibility for presentation of the message rests with the receiver manufacturer (who should tailor the presentation to suit the display provided), a number of control characters may be included within the text. They have been defined to have specific meanings, which manufacturers may choose to adopt if desired. As noted above, the 0D hex code is used to indicate the end of a message. Additionally, the code 0A hex is used to indicate a preferred line break. This marker indicates positions in the message where the author of the text would prefer a line break to occur in order to structure the text in a particular way. It is primarily useful for multiline displays. It may stand in place of a space character in the text string. Receivers choosing not to use control code or 0A hex should substitute a space at the position of the control code when displaying the message. A space should be substituted by the receiver display for any other unrecognised character.

Of course, receiver displays vary enormously, ranging from a single eight-character display on many RDS receivers to the display possibilities afforded by PC monitors or LCD screens. Because of the wide range of display types possible, it is up to the receiver manufacturer to develop software to optimise the presentation for the particular display chosen. For example, multiline displays should use normal word processing techniques to provide word wrap (to prevent breaking a word wherever possible). Such displays will need to accommodate more than 64 characters to allow for word wrapping and still show the message completely.

Similarly, displays with less than 64 characters, which rely upon scrolling the text to show the complete message, should apply intelligent routines to present the message in the best possible way. For example, receivers could remove unnecessary spaces inserted by the broadcaster as padding before display to prevent users scrolling through strings of blank spaces. If the resultant text length is shorter than or equal to the number of characters in the display, the display could show the message statically, without scrolling.

A RadioText message should be received in full *before* it is displayed. A receiver will know when a message is complete because it will be terminated by an 0D hex code or all 16 segments will have been sent. When a change in the Text A/Text B (new message) flag is detected, the receiver should begin to acquire the new message, which, when complete, should replace the former message, erasing it *entirely* from the display.

The RadioText feature can provide a wide range of additional information to the listener, including radio station information, such as telephone numbers, and programme-related information such as the CD numbers of the music being played, as shown in Figure 4.5. As a result, the safety issue of car receivers must be carefully considered, and some manufacturers are beginning to offer RT displays in car receivers. In this case, it is particularly important that users in vehicles are not distracted by RadioText messages. For example, users are permitted to switch RadioText deliberately "on" while "off" should be the default mode.

4.5 Decoder Identification (DI) and Programme TYpe Identification (PTYI)

The DI feature is used to indicate to the RDS receiver different operating modes to switch individual decoders on or off. The dynamic PTY indicator feature is used to indicate if PTY codes in the transmission are dynamically switched by the broadcaster. These two features are suitable for PTY-STANDBY and PTY-STORE modes of operation. (see Section 4.2).

DI is coded using a total of four bits, sent one bit per group over a *four group sequence* carried in type 0A, 0B, and 15B groups, in the same positions in each type.

The DI segment bit itself is carried in block 2, and the segment address is shared between PS and DI in two bits, as shown in Figure 4.6, to indicate which actual bit—d_0, d_1, d_2, or d_3—is being signalled in that particular group. At the recommended repetition rates, the status of each DI bit is signalled at

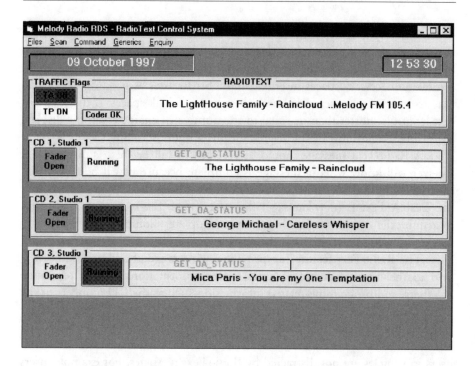

Figure 4.5 RadioText computer display: showing on-air RT messages and those cued for transmission. (*Source:* EBU.)

least once per second. In the case of type 15B groups, the segment address is solely used for DI addressing, as shown in Figure 4.7.

Table 4.3 shows the meaning of the DI bits. To date, bits d_0, d_1, and d_2 have not been popular with broadcasters. But some broadcasters have chosen to indicate that certain programmes (e.g., news) are signalled as mono by the d_0 bit, to allow a receiver to optimise audio performance.

However the RDS specification EN 50067:1998 now describes the use of the d_3 bit, which is assigned to signal static/dynamic PTY status. This helps an RDS receiver to assess the transmission and therefore determine how it will perform [1]. A broadcaster setting this bit to 1 is informing the receiver that the broadcaster has in place the infrastructure to manage and control the PTY codes broadcast on that and/or EON cross-referenced services. This will help prevent a receiver from remaining in PTY-STANDBY mode and staying indefinitely on a service that does not control PTY code switching. If d_3 is set to 0, the receiver should inform the listener that PTY-STANDBY mode is unavailable on that service—this can be done, for example, by displaying the information >No_Stby_<.

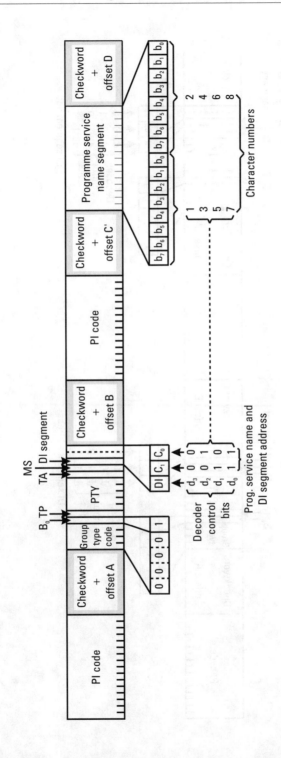

Figure 4.6 Type 0B groups, showing the DI segment and shared segment address bits. (*Source:* EBU.)

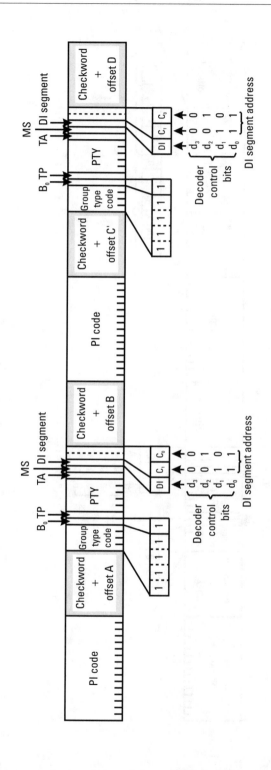

Figure 4.7 Type 15B groups, showing two DI segments and two segment address bits. (*Source:* EBU.)

Table 4.3
DI Bits: Meanings

Bit	Set to 0	Set to 1
d0	Mono	Stereo
d1	Not artificial head	Artificial head
d2	Not compressed	Compressed
d3	Static PTY	Dynamic PTY

Dynamic PTY indicates that the PTY on the tuned service or PTY referenced in type 14A groups, variant 13, is assessed at programme junctions and may be switched.

4.6 Programme Item Number (PIN)

The PIN feature is a machine-readable code assigned to each programme or programme item. Taken together with the PI code, it provides a unique reference for every programme. It may be used to automate receiver switch-on and/or recording.

PIN is encoded as 16 bits in block 4 of type 1 groups. The code may also be carried in variant 14 in the type 14A groups (EON) from transmitters cross-referencing this programme service. The PIN code expresses the published, scheduled start time of the programme, using the first five bits to represent the day of the month, the next five bits the hour, and the final six bits the minute, as shown in Figure 4.8.

The PIN feature should be transmitted at the rate of at least once per minute; however, when a programme number changes, the repetition rate of these groups (and the corresponding type 14A groups in other services) should be increased to transmit a burst of at least four groups in order to aid receiver responses.

So that the PIN may be used to properly control switch-on and automatic recording of programmes, the PIN code should accurately follow the on-air audio programme to take account of schedule changes or programmes overrunning their scheduled time. Valid PIN codes are those that have date information in the range 01–31, hours in the range 00–23, and minutes 00–59. Values outside the ranges indicate that no valid PIN code is being transmitted.

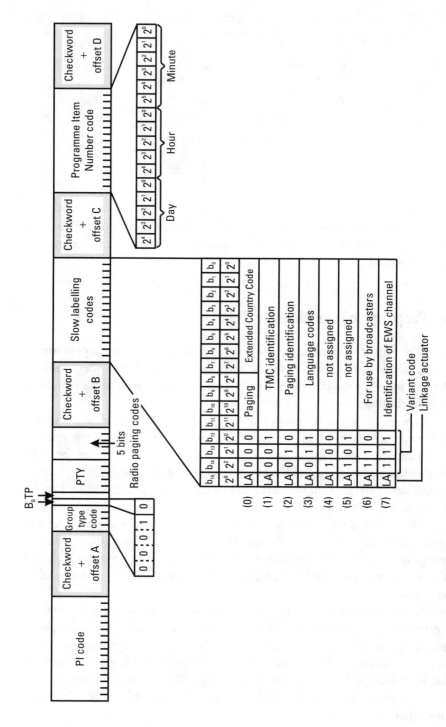

Figure 4.8 Type 1A group: showing the PIN feature in block 4. (*Source:* EBU.)

Although implementation of the PIN has generally been very slow to come to the market, it could be used as a feature that equates to radio on demand by allowing the listener to enter a PIN into an RDS receiver/cassette unit that could then record a programme for later recall. The PIN may be implemented in a receiver in a similar way to the PTY used in the PTY-STANDBY mode. In this case, it would be important that the scheduled start times published for programmes are presented consistently in all journals and programme guide listings so that the user can accurately enter programming, thus ensuring that the right program is recorded.

References

[1]   EN 50067:1998-Specification of the Radio Data System (RDS) for VHF/FM sound broadcasting in the frequency range from 87.5 to 108.0 MHz.

[2]   EIA/NAB-National Radio Systems Committee: United States RBDS Standard Version 2.0, Specification of the Radio Broadcast Data System (RBDS), 1997.

[3]   EBU /RDS Forum: BPN 009-Radio Data System (RDS) Guidelines, to be published in Autumn 1998.

5

Additional Information Features

5.1 Introduction

This chapter explores several optional RDS features originally developed to widen the information-carrying ability of the Radio Data System. The original developers realised that once an RDS receiver was designed with an eight-character display, some additional features could be added very economically, such as the Clock Time feature. RDS is also ideally suited for carrying other features requiring very low data capacity needs, such as the in-house and emergency warning systems, which are intended for special closed user group purposes.

However, the most important additional information feature, added to the RDS standard in 1992, is the EON feature, which has the amazing ability to keep an RDS receiver fully informed about other transmissions and help the receiver tune to those transmissions for services previously requested by the user.

5.2 Clock Time (CT)

The Clock Time feature is designed to enable a receiver to set and maintain accurate time and date, both for display and for control purposes. Clock Time is transmitted in a type 4A group using a total of 34 bits, as shown in Figure 5.1. A single 4A group is transmitted once per minute, inserted into the dynamic RDS multiplex data stream so that the minute edge will occur within ±0.1 seconds of the end of the Clock Time group. If transmitted, Clock Time

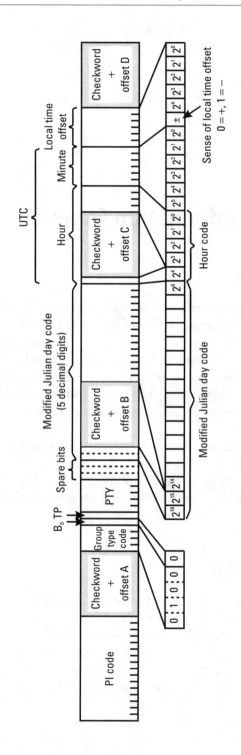

Figure 5.1 Type 4A group, showing the bits used for the CT feature. (*Source:* EBU.)

information must be accurately set *and* maintained. The CT information relates to the epoch immediately following the start of the next group.

Clock Time is expressed as coordinated universal time (UTC) plus local time offset. The local time offset is *expressed in multiples of half hours*, in the range −12 hours to +12 hours, and is coded as a six-bit binary number. The sense of the offset is indicated by a single bit, where /0/ = positive offset (east of zero degrees longitude) and /1/ = negative offset (west of zero degrees longitude). For example, 15.00 hours during the summer in the Central European time zone should be expressed as 13.00 hours (UTC) and have a positive offset of 4. Special care should be taken with the offset values when changing from daylight savings time (summer time) to winter time and vice versa to ensure the accuracy of transmitted data.

The date is expressed in terms of a modified Julian day. Note that the date changes at midnight UTC, not midnight local time. Conversion of time and date conventions are all described in Annex G of the RDS specification EN 50067:1998 [1]. It is noted that the formulae are applicable between March 1, 1900 to February 28, 2100.

As CT must be accurate, it is usual for a broadcaster to synchronise their CT transmissions to one of the national or international time references such as DCF (77.5 kHz) in Germany or MSF (60 kHz) in England. These time references are available over wide areas and relatively simple, highly reliable receivers can be installed wherever there is an RDS encoder—to feed it with an adequately accurate time code signal. When accurate time reference is lost, RDS encoders should automatically suppress transmission of type 4A groups—or they may set all bits to zero—until an accurate time has been reestablished.

Clock T is transmitted infrequently (only once per minute), and is not intended for direct display use by a receiver. The CT feature is used to set and update a free-running clock provided in the receiver. The receiver should usually present the user with a display of the local time by calculating it from the UTC data and the offset information. It is essential that the free-running clock continues to function without regular type 4A groups being received. For example, a user may choose to listen to a service not providing CT information, but he or she would still expect the clock to continue to display the correct time.

RDS receivers will often be operating in areas where reception of programme services in two different time zones is possible. Some portable receivers have implemented an update on/off switch to disable automatic updating of the clock by CT. This is done to avoid the clock showing different times, depending upon the tuned service.

The CT feature must be implemented to support some other RDS features. Both Radio Paging (see Chapter 8) and Traffic Message Channel (see Chapter 7) require CT as specified in EN 50067:1998, and some Open Data

Applications (ODAs, see Chapter 9) may also require CT to be implemented [1].

5.3 Enhanced Other Networks (EON)

The EON feature was designed to allow RDS to become "intelligent," especially from the perspective of large network broadcasters (e.g., the ARD companies of Germany, the BBC in the United Kingdom, and Swedish Radio) that each operate several national radio networks.

EON information allows the updating of a considerable number of features: AF, PIN, PS, PTY, TA flag, TP identification, and linkage information for programme services other than that of the currently tuned service. According to the features supported by the broadcaster and the RDS capacity available, some or all of this cross-reference information may be included. It is intended to allow the broadcaster to cross-reference information about *all* his or her services and, if mutual agreement can be reached, also about the services of *other* broadcasters.

EON information is transmitted in type 14 groups, which has 16 variants identified by four bits in block 2. The information itself is carried in block 3 of the group, and block 4 carries the PI code of the service to which the data in block 3 relates. This is shown in Figure 5.2. The variants may be transmitted in any order, but the repetition rate of these groups is arranged so that all data for all cross-referenced services is transmitted within a maximum of a 120-second cycle. Not more than 20 other services should be cross-referenced by any one service. This limit and the maximum 120-second cycle time are important factors in EON receiver design for achieving timely responses. In most areas, the total number of services available will not exceed 20. As a receiver changes location, the services within an RDS data stream may change, with new services being added and others removed; however, no more than 20 will ever be cross-referenced by EON from any one transmitter. By keeping note of the time since a service was referenced in EON, a receiver will be able to "mark for deletion" services that are no longer valid.

Each EON group is entirely self-contained so that the information can be used without reference to any other group.

Although type 14A groups are usually transmitted in a regular cycle, when the status of any information on a referenced service changes (e.g., PTY or PIN changes), three or four repetitions of each appropriate group type variant should be transmitted in rapid succession to assist an RDS receiver in responding quickly to the changed information, if required (e.g., for a receiver in the PTY-STANDBY mode).

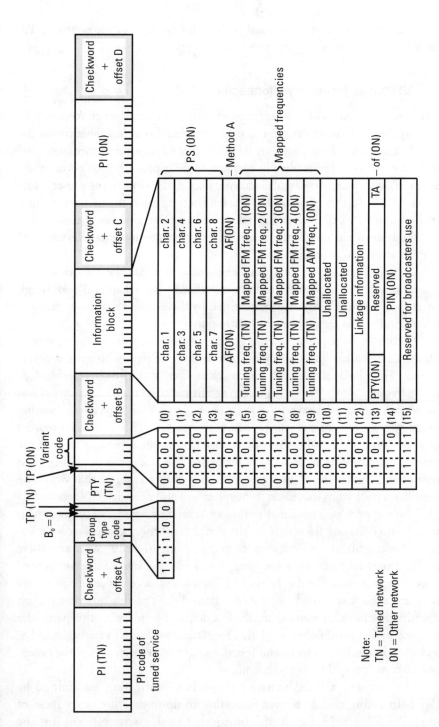

Figure 5.2 Type 14A groups: showing the many components contained. (*Source:* EBU.)

A different mechanism is used to indicate changes in the status of TA flags of referenced services—the type 14B group (see Chapter 6).

5.3.1 Alternative Frequency Information

Alternative Frequencies of cross-referenced services are transmitted to enable an EON-capable RDS receiver to build up a complete frequency table for all the services in memory. This allows an RDS receiver to respond immediately to a request for another service already memorised on a preset button, even if the user is well away from the original transmission used to source the preset information. It can do this because it will have been updated in the background, thus avoiding a PI search that may take a considerable amount of time in the increasingly crowded FM band. There are two transmission methods specified to convey the AF information.

Method-A (using variant 4) is identical in structure to the Method-A coding used in type 0A groups for tuned network AFs (see Chapter 3). Although theoretically the number of AFs transmitted by this method may be as many as 25, it is common to restrict the number to only a few.

The mapped-frequency method pairs the frequency of the tuned service with the corresponding frequency of the referenced service with similar coverage areas. The tuned service frequency is transmitted as the first byte in block 3, and the frequency to which it is paired is transmitted as the second byte. Each frequency in the tuned network AF list is paired in turn with the corresponding frequency for every other service being referenced. This method provides enhanced performance whereby a broadcaster transmits several services at the same powers, from the same transmitter masts. If a receiver is providing good reception of a service from one transmitter, it is usual to assume that switching to another service from that same transmitter will also provide good reception.

The mapped-frequency method uses variant 5 of type 14A groups to map one referenced service frequency to the tuned frequency. Occasionally, a frequency for the tuned network may be reused some distance away, and there may be more than one corresponding frequency for one of the referenced services. In this case, for the pairing of the second occurrence of the tuned frequency, variant 6 is used instead of variant 5. Use of this different variant explicitly indicates that more than one frequency is paired with this particular frequency in the tuned network AF list. Similarly, two further variants, 7 and 8, may be used to indicate a third and fourth frequency, respectively, in the exceptional circumstances that they are required.

The purpose of the Alternative Frequency information transmitted by EON is to ensure that a receiver maintains an up-to-date list or database of appropriate frequencies for all of a broadcaster's other services—not just the currently tuned one. This ensures that a receiver can switch quickly to the optimum frequency when a listener changes the preset selection. Figure 5.3 shows a conceptual model RDS receiver with a preset button memory array and additional "pool stores," which together use AF information derived from type 0A and type 14A groups. The PI codes and frequencies used are fictional but based on services available in the London area.

In the case of AFs transmitted by the mapped-frequency method, each frequency for a service is paired with one in the AF list of the tuned service. Because the receiver is presumably using the optimum frequency for the tuned service, it already "knows" the optimum frequency to use for the other service, should the listener choose to select the preset for it. In the case of AFs delivered by Method-A, although no direct correspondence exists, the receiver has only to quickly evaluate the AF list. Method-A lists delivered by EON are likely to be short, with no more than four or five frequencies, so evaluation of these frequencies can be almost instantaneous.

5.3.2 PIN and PTY Information

It is necessary to provide an EON-capable receiver with other information to enhance the performance; the "intelligence" that EON brings and which can be used to allow current listening to be interrupted by information or programmes from other services.

PTY is signalled in type 14A group, variant 13, and PIN in variant 14. Changes to either are indicated by immediately inserting three or four of the appropriate variants into the regular cycle of type 14A groups. This assists the receiver in responding promptly to changes, particularly when reception conditions are not perfect, which is usually the case since the repeated variants of type 14A groups increase the chance of "first time" decoding of the changed status of the feature.

When a receiver is in either PTY-STANDBY or PIN-STANDBY mode, a similar routine to that described above for EON traffic announcements should be followed (see Chapter 6). Switching to a programme should only occur at the start of a programme. Because the start (unlike traffic announcements) is not signalled by using a different group type, the start can only be reliably detected by monitoring and responding to a transition of PTY code or PIN received in type 14A groups.

5.3.3 PS Information

Perhaps the simplest EON feature is the inclusion of the PS feature carried in type 14A groups, variants 0 to 3, and the coding follows closely that used in the

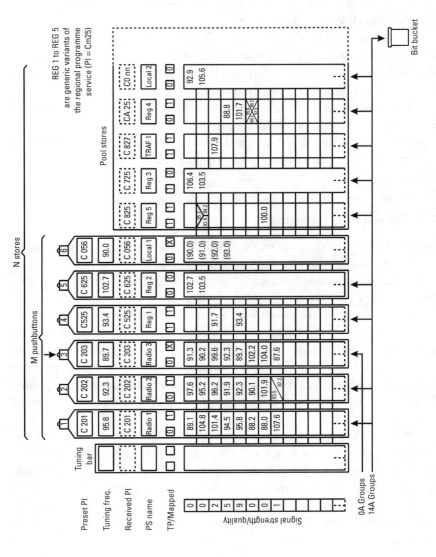

Figure 5.3 Conceptual radio with preset button and "pool store" memory. (*Source:* EBU.)

main PS feature described in Chapter 3. The eight characters are carried two at a time in these variants, so their display position is known to a receiver and they can be built up over the 120-second cycle period, to be placed in memory associated with any presets for instant recall.

5.3.4 TP/TA Information

Type 14A groups carry information in block 2 about TP flags of cross-referenced services in every group. Additionally, the type 14B group carries information about TA flags for cross-referenced services. Their use is described fully in Chapter 6.

5.4 In-House (IH) and Transparent Data Channel (TDC)

RDS can be used for completely nonprogramme-related data services, and these have been specified with the intention that consumer receivers do *not* respond in any way to their groups.

The IH data feature is reserved for data to be used solely by the service provider and/or transmission operator for their own applications.

In-House data is transmitted in the 37 bits reserved in the type 6A and 6B groups, as shown in Figure 5.4. The content of these bits is defined unilaterally by the service provider and/or transmission operator. The repetition rate is chosen to suit the application and available channel capacity.

Now that the RDS specification includes the ODA feature, the introduction of new data applications previously intended for IH or TDC should be considered carefully for registration as ODAs in preference to using these features, because the signalling associated with ODA is more flexible [1].

The Transparent Data Channel feature allows data to be sent to any of 32 addresses. The data is essentially free-format and may be for display, control, or computer applications.

TDC is transmitted in type 5 groups, as shown in Figure 5.5. Five bits in block 2 are used for up to 32 addresses. The data is carried as 16 bits in block 3 (type 5A group only) and 16 bits in block 4. The repetition rate for these groups must be chosen to suit the application, subject to maintaining a minimum rate for the basic RDS features.

5.5 Emergency Warning System (EWS)

The Emergency Warning System has been designed to take advantage of RDS FM transmissions—which already cover large broadcast service areas—to

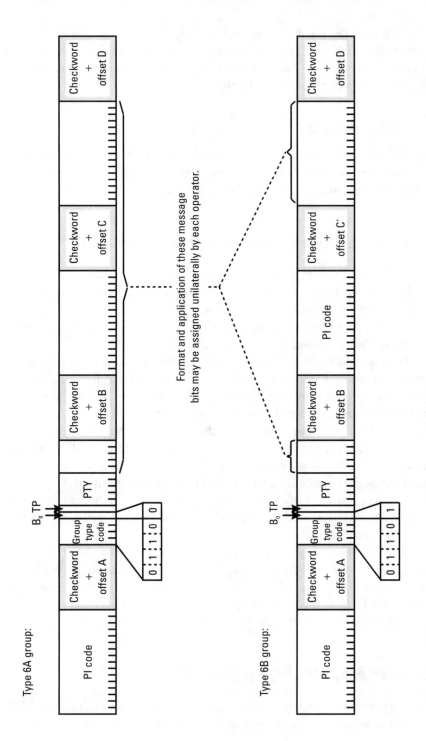

Figure 5.4 Types 6A and 6B groups, used for IH data. (*Source*: EBU.)

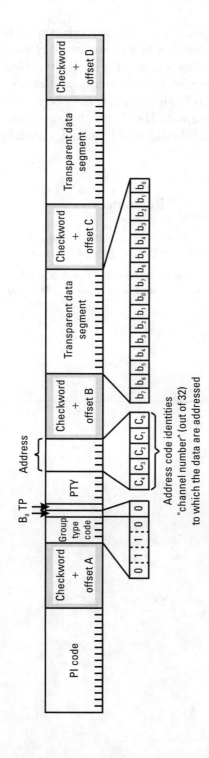

Figure 5.5 Type 5A group used for TDC, showing the addressing bits and data bits. (*Source:* EBU.)

occasionally convey emergency status data to specialised receivers. EWS data is carried in type 9A groups, which will be transmitted very infrequently unless an emergency or test transmission is required. The 37 bits in blocks 2, 3, and 4 are available for unilateral allocation in each country, as shown in Figure 5.6.

If the EWS feature is implemented, it must be signalled (say, once per minute) using type 1A groups, variant 7, and the transmission must also have the ECC feature implemented using type 1A groups, variant 0; these are shown in Figure 3.9.

References

[1]   EN 50067:1998-Specification of the Radio Data System (RDS) for VHF/FM sound broadcasting in the frequency range from 87.5 to 108.0 MHz.

[2]   EIA/NAB-National Radio Systems Committee: United States RBDS Standard version 2.0, Specification of the Radio Broadcast Data System (RBDS), 1997.

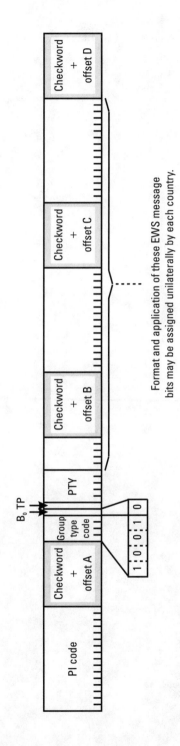

Figure 5.6   Type 9A groups used for EWS, showing the 37 available data bits. (*Source:* EBU.)

6

Traffic Information Services

6.1 Introduction

This chapter brings together the features needed to provide both simple and more comprehensive RDS-based traffic services. It was noted in Chapter 1 that the TP and TA flags have their origins in an earlier signalling system known as Autofahrer Rundfunk Information (ARI). In essence, it was designed for FM broadcasting using a 57-kHz subcarrier to signal just two parameters, which are directly modelled by the TP and TA features in the RDS specification. Additionally, it included the ability to indicate regionalisation, which RDS provides in the PI feature [1].

The fundamental aspects of the TP feature, which is used as a key tuning feature, are explained fully in Chapter 3. In this chapter, the use of both traffic service flags, TP and TA, together is explored. Then, an improved application of the RDS traffic service is described using the enhanced other networks (EON) feature in conjunction with the TP and TA flags by using a driver scenario. This scenario covers the circumstances of generating and transmitting all the RDS features from several radio stations to make up an overall RDS traffic service for the user.

6.2 RDS Traffic Services: Using the TP/TA Features

The RDS traffic service, which uses both TP and TA flags, is designed to allow listeners to receive and hear traffic reports even if listening to a CD/cassette, or with the receiver volume turned down or muted. Some receivers make use of the flags to make an unattended recording of the traffic information, so it is ready for replay by the driver upon request.

The TP flag is carried in block 2 of *every* group type and in block 4 of type 15B groups. The TA flag is carried in block 2 of type 0A and 0B groups; additionally, the TA flag is carried in blocks 2 and 4 of type 15B groups. (See Chapter 3, Figures 3.1 and 3.2 and Chapter 4, Figure 4.1) The use of these flags, so-called because they can be set to either /0/ or to /1/, is shown in Table 6.1.

Broadcasters must *not* set the TP flag to /1/, unless the TA flag is *dynamically* controlled. Furthermore, broadcasters must only set the TA flag to /1/ while a traffic announcement is in progress, although it may be necessary for transmission operational reasons to set it a few seconds (perhaps no more than three seconds) before the announcement starts. Delays inevitably occur in both the broadcast infrastructure *and* in RDS receiver processing, such that up to about two seconds may elapse between a studio setting the TA flag and a receiver responding as described above. The traffic service identification jingle, which is played from a source that can be used to trigger the mechanism, is often used to provide the very short advance time required to achieve the flag setting.

Even more importantly, broadcasters should ensure that mechanisms are in place to monitor the status of the TA flag, to assist in resetting it to /0/ immediately after traffic announcements have ended. This is very important to ensure that RDS receivers, having been automatically controlled by the broadcaster, are able to return to their previous *listener* setting (such as CD listening).

When tuned to a frequency with an RDS data stream indicating that both TP = /1/ and TA = /1/, a traffic announcement *is* in progress, so an RDS receiver should pause CD/cassette listening and revert to tuner listening. It must also increase volume, if necessary, to a predetermined volume level to ensure the announcement is audible. This receiver aspect is easiest to understand as follows: the volume level setting is automatic in the sense that it will be set at whatever level it was set at while listening to a previous announcement.

Table 6.1
TP and TA Flags Meanings

TP	TA	Meaning
0	0	No traffic service possible at all on or via this service
0	1	Traffic service provided via one or more EON cross-referenced services
1	0	Traffic service provided on this service and also possibly via EON services
1	1	Traffic report in progress on this service

Users can then adjust the volume level for their normal traffic announcement listening requirements, which may be different from other listening needs.

6.3 RDS Traffic Services: Using the EON and TP/TA Features

The EON feature opens the opportunity to large network broadcasters to signal to RDS receivers that traffic announcements are available and may be receivable, and to provide the listener with these announcements, even though they are being transmitted on *another* frequency. The basic function of EON, described in Chapter 5, is to build up a database of information about other transmissions (over a 120-second period) and, if appropriate, to place it into receiver memory. In the case of the traffic service, EON is used to provide dynamic information to an RDS receiver so that it can act very quickly to retune to a specific frequency; it knows from the database where a traffic announcement is taking place.

Type 14B groups are transmitted only when there is a change in status of the TA flag of the referenced service. This is signalled by a minimum of four (and up to eight) type 14B groups transmitted within two seconds, with the objective of causing rapid switching by the receiver to a traffic announcement on another service. In order to improve reception reliability, these groups should be spread over the two-second period in order to avoid a contiguous burst of groups being lost due to multipath propagation. Figure 6.1 shows all the TP and TA flags in both the type 14A and 14B groups.

6.3.1 A Traffic Event Scenario

The EON traffic service therefore uses quite a complex set of interactions between information being transmitted in the background, which will take some time for an RDS receiver to acquire, and information that must be acted upon immediately. The following explanation of these interactions is based on the situation found in the southeast of England. In this area, the BBC uses a combination of national and local services to deliver the necessary data for the RDS receiver and the local services are used to actually make traffic announcements on a regionalised basis. These transmissions cover county-based areas (a county is an administrative area that can normally be traversed in about 60 minutes of driving time). This arrangement offers relatively locally focussed information delivered by traffic presenters who have good knowledge about the area they are covering. This explanation is fictional but uses a broadly realistic idea to describe the interactions and includes some clever strategies used by the broadcaster to achieve an economic signalling system [2].

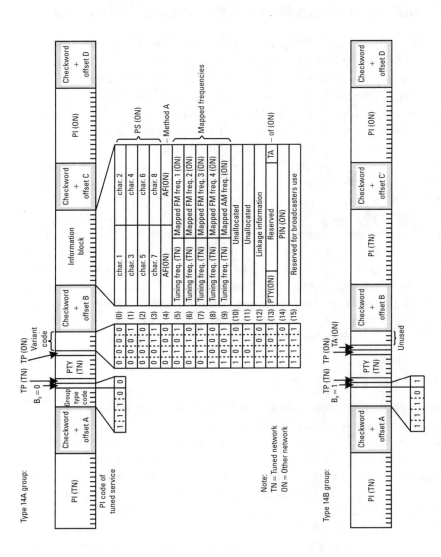

Figure 6.1 Type 14A and B groups: showing the all the traffic service flags. (*Source:* EBU.)

Figure 6.2 shows the scene in the southeast of England with a driver leaving Folkstone, near Dover, and travelling towards London via Maidstone. Our driver (1) is using his EON-capable RDS receiver to listen to a BBC service from the Dover transmitter (2). This transmitter actually transmits R2, R4, and Radio Kent. Our driver is now listening to BBC R4—the national news and current affairs network, which has transmitters all over the U.K.

The driver plans to use the Motorway M20 for his route, but an accident occurs on the M20 between Folkstone and Maidstone before he leaves the Folkstone urban area. He needs to know about this problem rather quickly in order to replan his route and take another road towards Maidstone. An event of this sort will quickly be attended to by the local traffic police patrol in a car that allows them to be in radio contact with their control room (3). The police will radio details about the event to the control room and, in this case, they will have to shut the motorway in the London-bound direction until the accident can be safely cleared.

The police control room will have a busy time managing an incident of this sort, but they realise that the various traffic and travel information (TTI) service providers need to be informed quickly about the event. Depending on the time of day, radio stations in the area use different TTI service providers,

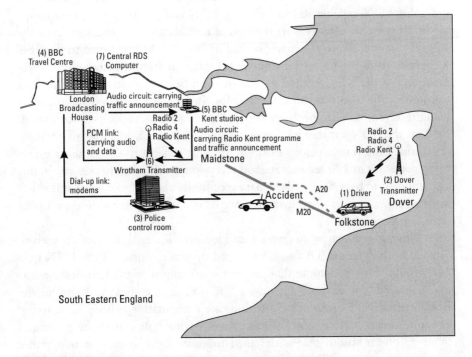

Figure 6.2 RDS EON traffic service in operation. (*Source:* EBU.)

including the BBC Travel Centre (4), based at Broadcasting House in London, and the Automobile Association Roadwatch service. In order to keep this example clear, it is assumed that the BBC Travel Centre is informed about the accident by the police control room. The information may be sent to the traffic presenter by dial-up modem link, fax, or a phone call, but this will depend upon operational circumstances.

So, now the BBC Travel Centre is aware of the accident and they will talk to the Radio Kent on-air studio to agree on the timing for a suitable traffic announcement to be broadcast. This announcement may be held for a few minutes until a suitable programme break occurs and, until a scheduled traffic announcement time slot comes up, or it may be given urgent status. In any case, the Radio Kent presenter will hand over control to the traffic presenter, who then can give the traffic announcement to Radio Kent listeners. It is already known that there is a Radio Kent transmitter near Dover, and there are others around the county to ensure good coverage, so the driver should be within range of a transmitter that will carry this announcement.

When a traffic announcement is being made with RDS implemented, the Radio Kent transmitters must have their TA flag set to /1/, but this will only help the user who is already tuned to a Radio Kent transmission, as described in Section 6.2. In this case, the driver is listening to Radio 4. However, he had preset his EON-capable RDS receiver to EON traffic service reception mode, so the receiver has already started to build up a database in memory from the BBC R4 service. When the driver started his journey, he was tuned to the BBC R4 service from the Dover transmitter (2). He may have already unknowingly made use of the RDS AF and PI features and been retuned to the Wrotham transmitter (6), also transmitting BBC R4. It does not matter which signal is being received, because both are carrying the background EON information. By now, the EON-capable RDS receiver has acquired considerable information about the BBC transmissions in the Kent area; indeed, it would know everything necessary in 120 seconds if good reception conditions are present. In this case, it needs to have acquired the data specifically about Radio Kent and either updated a preset button-related memory or placed the information in a pool store (see Figure 5.3).

Now, it is necessary to go back and look at the signalling systems used to provide both dynamic TA flag control and dynamic control of the EON type 14B groups. Firstly, assume that just as the starting traffic "identification jingle" is played, the Radio Kent on-air studio sends a command signal to the BBC central RDS computer (7) to indicate the situation: "BBC Kent traffic announcement starting now." (A message of this type has to be generated by every on-air studio that is designed to control the system so that traffic announcements can still be voiced by service providers from outside the BBC, who are feeding their audio into BBC on-air studios.) The central RDS computer receives this message and places it into a complete database of all BBC RDS data. From this database, the computer is able to generate appropriate update messages for RDS encoders wherever they may be. This is done by a suitable addressing system (see Chapter 11 for more details of studio to encoder communications and the UECP). These update messages are sent to all possible transmitters, including the Wrotham transmitter site in North Kent (6).

6.3.2 Clever Signalling

At this point, it is desirable to take a short diversion from the story and discuss the data circuits that are used, because this is where a clever use of RDS comes in. Between BBC Broadcasting House and the main transmitter sites such as Wrotham (6), there are wideband digital distribution circuits capable of carrying many audio channels and data. Thus, the Wrotham site has transmitters for the national services, fed by high-quality audio from London, and dynamic RDS is directly controlled from the central RDS computer. But the BBC Radio Kent transmitters at Wrotham and Dover, giving the county coverage across Kent, do *not* have direct control from the central RDS computer. For example, the Dover transmitter receives RDS update data from off-air RDS signals by using an off-air receiver tuned to BBC R2 and then decoding the EON data about Radio Kent. The RDS decoder is connected to the RDS encoder, and this in turn is commanded to change the TA flag on the local Radio Kent transmission. Thus, the cost of data circuits is greatly reduced by this on-air signalling technique.

6.3.3 Update Messages Content

Now, consider the update messages and their content. The update command from the central RDS computer instructs encoders referencing Radio Kent as an "other network" to set the TA flag in Radio Kent EON groups, which results in the required transmission of up to eight type 14B groups within two seconds. This data is carried on-air to the Radio Kent transmitter at Dover, of course, and it is also transmitted on BBC R4 from Wrotham (6). In addition, it is transmitted on BBC R4 from Dover (2), which is a rebroadcast transmitter.

As a result, the driver whose EON-capable RDS receiver is tuned to BBC R4 somewhere in the Kent area and receiving an RDS data stream with EON information receives a specific vector command. The type 14B groups vector a receiver to a particular PI code, probably held in memory; if it is memorised, either in preset button memory or in a pool store memory, then associated alternative frequencies (AFs) will already have been noted for particular signal strength/quality. Thus, appropriate frequencies for Radio Kent are already known, and the EON receiver can just check the known frequencies for suitable signal strength and switch to the traffic announcement if the PI code match is satisfied. Once tuned directly to a Radio Kent transmitter for a traffic announcement, the receiver monitors the TA flag in the type 0 and 15B groups until it is set to /0/, whereupon it retunes to the original frequency and volume level setting.

Thus, the driver is able to hear the traffic announcement from BBC Radio Kent even though he had been tuned to BBC R4, and after the announcement the receiver is retuned back to BBC R4. He heard about the accident and diverted onto the A20 main road!

6.3.4 Receiver Reactions

An EON-capable RDS receiver will be controlled to give the listener a requested response by commands from the broadcaster. So, it must be certain that it does this well, and a number of steps should be followed when set into the EON traffic mode.

Cross-referenced services that provide traffic announcements are indicated as such by setting the TP (ON) bit (in block 2) to /1/ in the tuned service type 14A groups. This bit identifies the cross-referenced service as a traffic announcement service to the receiver, the purpose of which is to ensure that data for this service is stored in receiver memory prior to it being required for a traffic announcement.

The traffic announcement is on the service referenced by the PI code in block 4 of the type 14B group. The receiver should check the signal strength of the stored frequency for that service. If unacceptable, it should check, in turn, any others in the AF list for that PI and tune to the best frequency, if any others are receivable in that area. It should also check that the PI code matches that in block 4 of the type 14 B group, and then wait for a few seconds to establish that the TA flag has been set to /1/.

When a traffic announcement takes place, the broadcaster sends up to eight type 14B groups with the TA (ON) bit set to /1/ in rapid succession. Although a receiver only requires a single type 14B group to switch to a traffic announcement on the service indicated by the PI code in block 4, sending several groups increases the likelihood of detection by the receiver. To allow the receiver some response time to switch, these type 14 B groups should begin a few seconds prior to the start of the traffic announcement's spoken content. Of course, the TA flag in type 0A (or 0B and 15B) groups of the service providing the traffic announcement must also be set.

At the end of the traffic announcement, the TA flag in the traffic announcement providing service must be promptly cleared (reset to /0/). Optionally, type 14B groups with TA (ON) set to /0/ may be sent. They are *only* used by the broadcaster for internal network management purposes, which may be part of the complex scenario painted before to reduce the number of dedicated circuits within the transmission systems, to facilitate all the commands required.

References

[1]   CENELEC, EN 50067:1998-Specifications of the Radio Data System (RDS.) for VHF/FM broadcasting, European Committee for Electrical Standardisation (CENELEC), 35B rue de Stassart, B-1050 Brussels, April 1998.

[2]   Howe, P., "RDS-The Engineering Concept," BBC Radio, 1990

7

Intelligent Transport Systems and RDS-TMC

7.1 Introduction

Since the European Commission has chosen the RDS-TMC as a priority technology to be introduced for the provision of traffic and travel information on the major roads within the European Union member states, those involved in transport telematics have given much attention to RDS. In this chapter, we explain the strategic and policy issues that surround RDS-TMC—the market trends for telematics equipment in general and RDS-TMC in particular. The resulting human/machine interface problems and safety aspects that this poses are also covered.

The objectives to be achieved, and the history of RDS-TMC development (which started some 15 years ago), are analysed. We can see that in spite of this long development effort, RDS-TMC is not yet meeting the expectations that the politicians had for this technology.

The standardisation and institutional implementation issues, together with the basic principles of RDS-TMC coding, are explained, with some examples that help to better understand the technical detail of how RDS-TMC works. We show aspects of existing RDS-TMC receivers and where the problems are in their implementation.

Finally, we refer to alternative technologies and services that make it possible to see RDS-TMC in a much wider context—that it is in a long-term development leading to a significant improvement of the infrastructures required to provide better information services to the mobile road users.

7.2 Strategic and Policy Issues

In Europe, Japan, and the United States, huge transport telematics technology research and development programmes have been underway for several years. Their main focus is to coordinate the deployment of intelligent transport systems, which requires that a certain number of technical standards be developed. On the European level, this is undertaken by CEN (TC 278), CENELEC, and ETSI, and on a world level by the International Standards Organisation (ISO, TC 204), IEC, and the International Telecommunication Union (ITU). The term "telematics" is derived from the terms "telecommunications" and "informatics," and describes the fact that information processed by means of computers is being distributed to assist users to better cope with different transport modes for which the complexity, interdependence, and density is ever increasing. In other words, information and communication technologies are increasingly used to improve the delivery of general interest services [1]. Figure 7.1 shows one view of the telematics scene.

Traffic and travel information (TTI) is one of the general interest services that is being developed under these programmes. The use of RDS technology to deliver TTI messages within the European Union has long been recognised by the European Commission. This is in line with the resolutions of the European Conference of Ministers of Transport (ECMT) and the Council of the European Union, who have recommended, as a priority action, since 1995, the introduction of a new language-independent TTI service called RDS-TMC, where TMC means Traffic Message Channel [2, 3].

7.3 Market Trends for Telematics Terminal Equipment

The equipment we are particularly interested in is the RDS-TMC receiver—as part of a car radio designed to present improved traffic and travel information services to the end user. We are also interested in navigational units that can be autonomous or connected (also called dual-mode) to information services operated via mobile telephones (GSM), or RDS-TMC receivers for the dynamic updating of the route guidance information used on a particular trip.

The market potential for navigational systems can simply be considered as gigantic, and the evolution becomes obvious from the figures published in [5] and reproduced in Table 7.1.

We can note several trends. In Japan, this market is already very big, but the European and U.S. markets are following. In Europe today, the biggest market is Germany where two-thirds of all upper-class car models are already factory-fitted with navigational systems.

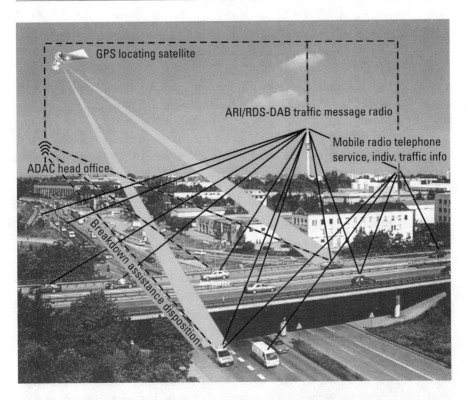

Figure 7.1 The German automobile club ADAC's view of a number of upcoming transport telematics road information services, among which is RDS-TMC. (*Source:* ADAC.)

Table 7.1
1995–2000, Number of Annually Sold Car Navigational Units

	1995	1996	1997	1998	1999	2000
In Europe	12,000	50,000	155,000	350,000	700,000	1,000,000
In Germany	12,000	40,000	110,000	200,000	320,000	500,000
In USA	5,000	20,000	200,000	500,000	—	—
In Japan	600,000	800,000	1,000,000	1,200,000	1,500,000	—

In 1997/98 alone in Japan, many of the navigational systems sold already had functionality for dynamic updating via the Vehicle Information and Communication System (VICS) using DARC FM-multiplex data broadcast technology. In Europe, similar systems first came onto the market in 1997, using

RDS-TMC for the dynamic updating; however, GSM and digital audio broadcasting (DAB) are likely to be used as well before the year 2000.

We can also note another trend. In the past, traffic information receivers were essentially car radios for which there was a significant aftermarket—supported by more than 50 different consumer electronic manufacturers, mainly from Europe and the Far East, supplying a wide range of products.

Since RDS has been on the market, car manufacturers have tended more and more to integrate the car radio in conjunction with multifunctional displays on the dashboard, and these will be increasingly used together with other telematics functionality. Additionally, they order original equipment manufacturer (OEM) equipment meeting their own interface requirements, opposing any standardisation of these interfaces.

Most recently, the German mobile telephone provider Mannesmann-Autocom acquired VDO, a large company that manufactures vehicle instruments and car dashboards. Then it acquired Philips Car Systems, one of the major car radio manufacturers and also an important supplier of telematic terminals and navigational systems. All this indicates that the aftermarket for all these products may gradually shrink and that a number of different telematic systems, as accepted by the market, will coexist through telematics equipment that is delivered with the car to the end user. On the other hand, due to the rapid progress in telematics technology, it is reasonable to assume that the predicted lifetime of the terminal equipment will be significantly shorter than that of a car (11.5 years, on average, in Germany). This trend opens new opportunities for the shrinking aftermarket.

In Germany in 1996, 60% of the three million GSM telephones already in use were, according to a study made by the German automobile industry association VDA, in one way or another also used in cars, which means that 1.7 million cars in Germany are equipped with mobile communication terminals. Some 96% of the units sold in 1996 were handheld units, of which 70% were supplied with optional equipment for fixed car installation. There is also a trend to integrate a GSM telephone with the car radio [6].

In 1994, the TELTEN project of the European Commission estimated the probable market penetration for RDS-TMC receivers for two scenarios: an optimistic one and a pessimistic one. Estimated market penetration in motor vehicles within Europe for RDS-TMC receivers was, in the optimistic case, 2% in 1998, 10% in 2000, and 61% in 2005. In the pessimistic case, estimates were 1% in 1998, 5% in 2000 and 48% in 2005. The latter, translated into absolute figures, predicts for RDS-TMC a European market of 31,000 units in 1998, 235,000 units in the year 2000 and 2.3 million units in the year 2005 [7].

7.4 Safety Aspects of Presentation of Traffic and Travel Information in Moving Vehicles

As more and more in-vehicle services come to the market, equipment add-ons might not be as thoroughly researched as equipment fitted in factories. There is also reason for concern that multiple in-vehicle services might jeopardise safety. There can be many different devices that communicate, and that in turn will present new information to the driver, distracting attention from the traffic around the vehicle on the road. For example, using a handheld GSM telephone while driving a car is generally considered to be dangerous. A study conducted in 1997 for the German automobile club ADAC shows that even using a hands-free fixed telephone distracts a lot the driver's attention from the traffic, and statistics about driver behaviour have indicated that all sorts of mistakes are being caused by distracted drivers [8].

As we have seen, telephones, navigational devices, and RDS-TMC units will be increasingly used. In such a case, designers will have to give particular attention to human/machine interface aspects. It will be necessary to integrate several communication functions into the same unit; also, the same display in the dashboard for presenting the driver with stress-free information will be used. Driver information will be presented in sequence, and not simultaneously, about several of these functions: GSM information services to be called up or based on automatic call-back, navigational information, and RDS-TMC message presentation.

More and more types of in-vehicle equipment thus require interoperability studies to be conducted so that guidelines can be established to reduce stress and distraction to the motorist when driving. Although the ECMT and the EC are fairly active in researching these issues, clear European guidelines on these safety issues and related certification and labelling mechanisms did not exist before 1998 for transport telematics terminal equipment. However, in the United Kingdom, there is already a draft for development prepared by the Department of Transport, and in Germany, the public and private sectors have agreed on a code of practice on the functions and security of in-vehicle systems [9–11]. During 1998, this matter has received urgent EC-supported investigation and has become a key research issue.

7.5   RDS-TMC

7.5.1   Objectives to be Achieved

The objective of RDS-TMC is to broadcast traffic and travel information (TTI) messages on VHF/FM broadcast transmissions using RDS. The messages are digitally coded in such a way that they are language-independent. The coding also permits users to receive only those messages that are relevant to their needs. Thus, it may be possible for a tourist to travel, for example, from London to Rome, through Belgium, Germany, Switzerland, France, and Italy, always getting the traffic announcements via RDS-TMC in his or her own language, the location codes being on a smart card that was purchased in London. But for this objective to be achieved, apart from a European standard specifying the system, a large number of harmonised administrative arrangements have to be agreed to among the countries concerned.

The TTI messages that are distributed will, of course, be conveyed to a variety of media that may include traffic announcements presented via speech synthesis in the language required. The messages will be filtered on the basis of criteria derived from the needs of the end user (e.g., location, direction, route). This filtering is aimed at enabling the user to receive only relevant information, selected from all the messages that are available on an RDS-TMC service at any given time.

The importance of filtering is demonstrated in this example. Given the maximum capacity of RDS-TMC, it will be possible to transmit about 300 different messages per hour. If the same messages were spoken and each only took 15 seconds, the total announcement time would be 75 minutes—an obvious information overload! In addition, RDS-TMC messages could be used to provide assistance for route planning, use of alternative transport facilities and routes, and dynamic updating of electronic maps and systems used for navigational aids.

In other words, RDS-TMC will be provided to simply give assistance to the question of how to get safely and efficiently from point A to point B.

7.5.2   History of the RDS-TMC Development

The possibility of using digitally coded traffic information within the RDS data stream was first mentioned by Bosch/Blaupunkt in 1984 at the Eurotravel conference in Grado, Italy. Bosch/Blaupunkt immediately started the technical development in Germany, and Philips in the Netherlands followed shortly afterwards. One year later, the European Association of Consumer Electronics Manufacturers, known then as Eurotech (now European Association of

Consumer Electronics Manufacturers (EACEM)), proposed to the European Broadcasting Union (EBU) to support the development of the TMC feature, called at that time Comprehensive Information for Motorists (CIM).

The EBU accepted this offer and created a group of RDS specialists, in which the interested manufacturers participated. In 1986, Philips and Blaupunkt submitted a common proposal that became the basis for further development of the TMC transmission protocol, defined in the EC/DRIVE I project as ALERT-C in 1991. A list of events was first issued by the EBU, and by 1988, it already covered seven languages. The list, now extended, is standardised by CEN as prENV 12313-2, but only officially existed in 1997 in the so-called CEN "Euro-English" version. The translations to be used in the various member states of the EU were still under development or, in 1997, available as a draft through the FORCE/ECORTIS projects of the EC.

In the years 1992–1994, a number of RDS-TMC field trials (14 Euro-regional projects in total) took place as part of the EC/DRIVE II programme on advanced transport telematics to test the feasibility of the service. The experience gained was largely positive, so that in September 1995, the Council of Ministers, meeting in Essen (Germany), recommended the general introduction of RDS-TMC in all EU member states.

In addition, in the summer of 1995, the EC Directorate DG VII committed itself to financially support an action plan covering the years 1995–1999 that will stimulate with "seed money" the introduction of RDS-TMC (see Table 7.2, which gives the total budgets for all RTI projects, including RDS-TMC). Since 1996, DG VII agreed to fund several RDS-TMC implementation initiatives, among them two European projects (FORCE/ECORTIS and EDEN) and several Euro-regional (VIKING, CENTRICO, CORVETTE, SERTI, and ARTS) projects, as well as national

Table 7.2
European Commission—DG VII Budget Line to Support RTI Implementation

	DG VII TEN Budget Line Support (in Million ECU)					
	TEN-T allocation	Scenario as input to discussions with Financial Assistance Committee				
	Decision 95	96	97	98	99	95–99
Total for 14 priority projects	182	210	268	324	366	1,350
Traffic management	45	54	67	82	92	340
Other important projects	13	16	17	24	38	108
TEN transport budget line	240	280	352	430	496	1,798

projects that all together involve Austria, Belgium, Denmark, Finland, France, Germany, Italy, the Netherlands, Portugal, Spain, Sweden, and the United Kingdom. This led to many regional RDS-TMC implementations on important sections of the Trans-European Road Network (TERN), concerning the main trunk roads in the in European Union.

The year 1997 was when the first RDS-TMC introductions started in a few EU member states (or parts of them), on a preoperational basis in most cases. The others have firm plans to follow this plan up to the year 2000. As a priority, these new RDS-TMC services intend first of all to cover all major roads belonging to the TERN.

Since a public and open service is generally expected to become available all over the European Union, it is hoped within the EC that public/private partnerships (i.e., partnerships between road authorities, police, automobile clubs, industry, broadcasters, etc.) will become possible within the EU member states, and that these states will then work together actively towards achieving RDS-TMC implementation, supporting the policy addressed in the resolution from the Council of EU Ministers.

7.5.3 The Pan-European Service Objective and the Memoranda of Understanding

The European Commission requested, in line with the European Council Resolutions of September 1995 and March 1997, that EU member states should promote the introduction of RDS-TMC and implement their services according to the proposed prestandards to ensure continuity and interoperability.

In the autumn of 1997, to better achieve this objective, the European Commission launched two Memoranda of Understanding (MoU), one encouraging the introduction of RDS-TMC services and the other recommending the DATEX protocol for communication between traffic information centres. Both were to be further developed and implemented following the recommendations of the EDEN project for transborder TTI data exchange.

Previous to these MoUs, the European Commission pursued the objective of providing to the people of Europe, a free Pan-European TTI service (PES) covering the traffic situation on the Trans-European Road Network (TERN) using RDS-TMC as the transmission mode. However, during the drafting of the MoUs, the term PES was dropped to favour a number of different but compatible services and without insisting that the services should be free to the end user on a European scale.

7.5.4 Institutional Challenges of RDS-TMC Service Provision

7.5.4.1 General Considerations

In order to achieve a European-wide service, a number of institutional agreements appear to be necessary, in particular on smart cards and location databases. There also needs to be coordination of the updating of these databases on the European scale, including the availability of the data to any service provider, the identification of the service, and the automated or assisted handover to a corresponding service in an adjacent region. Also, a European-wide service should be freely available everywhere, and if smart cards are used, they should be the same all over Europe—with a sufficient memory capacity so that a driver travelling large distances does not have to swap smart cards too often.

The TMC information to be distributed requires extensive data collection facilities to be installed, with traffic information centres to be created on a regional level and with a standardised data exchange protocol being used in operations between them. DATEX-NET is recommended, for which a standard is being elaborated on by CEN 278. This standard will be further developed through the EC-funded EDEN project.

Any RDS-TMC service provider should be given access to this data, so that they can offer a high-quality service from the end-user's point of view. When such data is not available because of a lack of infrastructure, which is still the case on many sectors of the TERN, the RDS-TMC service will not be very interesting for the end user. There are a number of private companies that are prepared to become involved in the data collection and operation of travel and traffic information centres, but they will have to recover their operational costs and would be reluctant to contribute to an open and freely available Pan-European service. The alternative, then, is data encryption and the provision of the service against payment of a subscription fee. The implementation and operation of this concept on a European scale appears to be impossible, however, because too many different interest groups are involved. A kind of patchwork of different RDS-TMC services is then more likely because of too little direction being given by the European Commission.

While public service broadcasters in the EBU are generally not against the formation of public and private sector partnerships, they won't, in the majority of cases, seek to become involved in the provision of an encrypted RDS-TMC data broadcast service.

In the following section, we describe in more detail the coding of RDS-TMC and the institutional requirements relating to the management of location databases and smart cards. Some of the models described were developed in 1996 for the European Commission. This was done within the EPISODE project that the EBU undertook in support of the broadcast sector with the view of assisting the implementation of RDS-TMC.

7.5.4.2 The Pan-European Management of TMC Location Databases

The location code database is central to the operation of any TMC system, in essence providing one of the two main keys to the encoding and decoding of data (the other being the event list). The following reviews the function of the database, identifies the roles and responsibilities of its owner, and makes recommendations that will enable the rapid development of a Pan-European system with fair competition among operators.

In any TMC system, it is necessary for the *same* location database information to be used at all traffic information centres (TICs), by all service providers (SPs), and by all devices that receive information from those sources. Such a system is depicted in Figure 7.2, which introduces the role of a card provider, translating the available database to a physical and marketable form. If any of the system elements have a database that differs from the others, performance will be less than optimal. For example, if the receiver were to have an old database missing new location codes, then messages relating to these locations could not be received and presented to the motorist. Such omission may be regarded as simply unfortunate, but could even have some safety implications.

Within the system described above, there can only be *one* database per area or country, and all system elements (TIC, SP, card provider, and user)

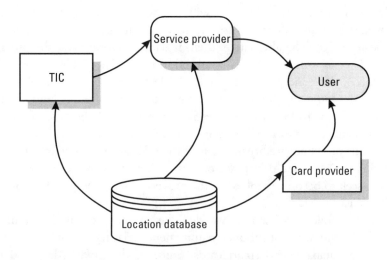

Figure 7.2 Location database influence diagram. (*Source:* EBU.)

must derive their individual databases from this one source. It is not possible for more than one location database to be used within this system, whereas it is possible for TICs, service providers, card providers, and receiver manufacturers to operate competitively. Clearly, the owner of the location database is, and must be, in a monopoly situation.

If fair competition and reasonable access is to be given to all operators, providers, and manufacturers, it is imperative that strict regulations and guidelines be imposed on database owners. We believe that such databases should be owned by nonprofit organisations, preferably by public authorities, and maintained by contractors. Costs may be offset by reasonable charges for acquisition of the database, which may be licensed for a period (to maintain system integrity) to operators, service providers, and card providers.

The task of the location database owner is to provide a fully maintained database, accessible on fair and equal terms to all who may request it and is updated at regular and declared intervals. In order to ensure that all system components are kept to the latest possible standard, it will be necessary for database owners to declare in advance when a new update is to be made available so that, for example, card providers may ensure that mass production is suitably synchronised.

The Pan-European vision of TMC foresees that a traveller can buy or rent a smart card in Italy that has location codes for London but with events pronounced in Italian. Such a card needs to be assembled using information derived from the U.K. database. This principle extends all over Europe, and it is clear that each database needs to be made very widely available.

Technical standards for information exchange need to be developed, such that any potential operator/supplier can obtain information from any database owner. Possibly the best way of achieving such interchangeability is to use open database communication (ODBC) protocols, which will permit operators to derive whole or defined subsets of a database at ease—for example, just locations in the northeast of France.

As memory sizes increase, card providers will wish to market products that cover larger geographical areas, and it will be possible and necessary to include locations derived from more than one location database. Clearly, there are distinct advantages in encouraging database owners to synchronise their updates so that it is possible to manufacture products containing the latest information from several sources. Such a situation is depicted in Figure 7.3.

With such an aim in mind, a nonprofit-making European database exchange organisation should be considered that coordinates and collects information from individual database owners and makes this available to all—subject to appropriate licences being issued.

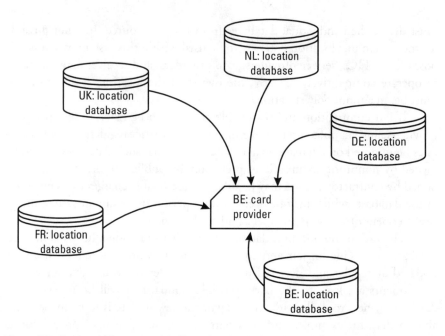

Figure 7.3 Card provider: example of location database information sources (partial example). (*Source:* EBU.)

7.5.4.3 Considerations Concerning the Use of Smart Cards and CD-ROMs as Storage Devices for the Location Codes

The access to the TTI messages will be enabled through an integrated circuit (IC) memory chip card using the ISO 7816 standard for the physical support (initially, 512-KB ROM). It will use a proprietary data recording and management format jointly developed by Bosch and Philips (also called the "TMC Chip Card," offered as one possible European industry standard in the so-called Blue Book, which is also an RDS-TMC technology licence package from Bosch/Philips to other receiver manufacturers). The TMC Chip Card is designed to contain appropriate location databases. The alternative solution to storing location databases on the TMC Chip Card is the use of a CD-ROM. Because no specification for recording the database has been internationally agreed to, manufacturers will have to use their own formats which, given the confusion that this will create for the end user, could well result in a rejection of RDS-TMC products by the consumer.

7.5.4.4 Pan-European Requirements for Smart Cards and Payment Methods

A smart card or CD-ROM is destined to become the means by which database information is loaded into the receiver, and by which it is updated as roads are improved. In the following, the function of these products is reviewed. The requirements for payment in a genuine Pan-European system are identified, and a simple and effective mechanism is proposed for deriving an income from the sale or rental of devices that may be distributed by or on behalf of service providers who cannot attract public support.

Each TMC receiver needs to have access to both a location code database and an event database coded in the user's language. Without these databases, the data stream arriving is meaningless. Each user will wish to acquire a series of databases, covering the areas in which he or she will travel. These are expected to be distributed in the form of a smart card, CD-ROM, or some other exchange media.

It is expected that smart cards or CD-ROMs will be supplied by a variety of providers, formatted to appropriate standards using data derived from national databases. In order to ensure that a high quality of service is maintained, it is imperative for these storage devices to have a version number and an expiration date, guaranteeing that out-of-date versions are no longer usable.

Each and every service provider will need to fund the costs of service provision in some way. In many countries, this will be by public grant. However, some countries are unwilling to support these services, and thus a mechanism must be included to generate a cash flow to pay for the service. This will enable a genuine Pan-European TMC service to be realised.

The simplest means of providing payment for a service is to derive income from the sale of cards to the user. These cards will need to be replaced periodically in order to ensure consistent service quality, and thus a source of regular income can be realised.

The profit on the sale or rental of cards may be used to finance the costs of obtaining the databases used, and to finance the service provider offering the data being delivered to the user.

It is expected that in many areas an RDS-TMC service may be available from more than one service provider (e.g., in the United Kingdom through both automobile clubs, the Automobile Association (AA), and the Royal Automobile Club (RAC)), permitting competition to push quality up and costs down. The RDS-TMC standards require broadcasting a service identification code to be within the data stream that will enable a receiver to locate the service operated by a preferred provider. To enable such a system to operate to the benefit of the consumer, cards need to have the ability to carry one or more service identification (SId) codes (rather in the same manner as a GSM card does today), allowing a TMC receiver to automatically find the required service(s). Relationships could be established between service and card providers, permitting one card to give access to more than one service, as shown in Figure 7.3.

Figure 7.4 shows an example of a card produced that offers access to the services of service providers SP 1 and SP 2, but not to those of SP 3. Service providers SP 2 and SP 3 are from one country, and SP 1is from another.

The card depicted in Figure 7.4 uses elements from the databases of both countries. It is expected that money to be allocated to the respective service providers could be distributed in proportion to the number of codes used from each country, the area of coverage, or some other method. In any case, the card provider is supplying a product that is licensed by both countries' location database owners and by both of the service providers.

7.5.5 RDS-TMC Standards

7.5.5.1 Introduction to RDS-TMC Standards

Because TMC services will be conveyed by the RDS subcarrier, and added to the baseband multiplex signal of FM Band II transmissions, it is usual to consider the RDS standard EN 50067:1998 as the core specification. This

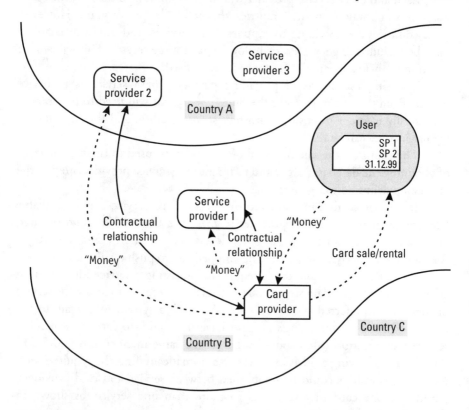

Figure 7.4 Service and card provider relationships and payment mechanisms. (*Source:* EBU.)

specification, when first published in 1990 and updated in 1992, already contained references to the RDS-TMC feature, mentioning both type 8A groups and a variant of the type 1A group that would be used for TMC identification.

However, the needs of a TMC service were not well defined at that stage, and work over recent years by CEN TC 278 SWG 4.1 and by the EC-funded FORCE Project has clarified the situation, resulting in the need to define more aspects of RDS-TMC in additional standards.

7.5.5.2 The Specification of the RDS System (CENELEC EN 50067:1998): the Core RDS Standard

The upgraded RDS standard EN 50067:1998 was developed by joint work in the RDS Forum and CENELEC TC 207, using a very wide range of experts. This version of the standard now fully includes the new open data application (ODA) feature (see Chapter 9), originally suggested for RDS-TMC purposes. It also includes specific references and "hooks" to ENV 12313-1: June 1997, where necessary. The former RDS-TMC references were kept to leave the door open for non-ODA implementations of RDS-TMC in the short term, but it is recommended that these implementations no longer are recommended to be maintained.

7.5.5.3 The Coding Protocol for RDS-TMC (ENV 12313-1): the So-Called ALERT-C Protocol

The coding protocol for RDS-TMC is defined in ENV 12313-1, which was accepted by voting within CEN in early 1996, but was the subject of an addendum and corrigendum. This addendum was originally issued in June 1996, mainly in response to comments from CEN TC 278. In view of this, ENV 12313-1, Version 2.20, dated September 1996 ,was redrafted to include all previous editorial changes noted in the addendum and corrigendum, and this has been subsequently further improved to become ENV 12313-1, Version 3.0, dated June 1997. Therefore, this should now be taken as the main standard describing the so-called, ALERT-C protocol. This document contains both the TMC message protocol and the RDS-TMC-related coding used by the RDS-TMC feature, using type (1A), 3A, and 8A groups to convey the TMC messages with event and location codes and network and service layer information.

7.5.5.4 Event and Information Codes for RDS-TMC (prENV 12313-2): the ALERT-C Event List

The event and information codes for RDS-TMC required by the ALERT-C protocol are specified in prENV 12313-2, FINAL DRAFT, dated May 1997. Although this standard does not actually require any further work to operate a service, it is seen as necessary to adopt versions for each European country/language. For example, in the United Kingdom, it is normal to describe distances in miles, but distance is described quite differently in other countries (in kilometres), so an agreed upon definition is useful to help uniform message generation and TMC decoder presentation.

7.5.5.5 Location Referencing Rules for RDS-TMC (prENV 278/7/3/0005 DRAFT)

Location referencing has been one of the most contentious issues, and rules for RDS-TMC are available as a draft dated March 1997. However, progress on CEN TC 278 SWG 7.3 has been rather slow and many workers are now aware that there needs to be considerable effort put into resolving the outstanding issues. In particular, location database development which had begun using a former draft has shown that considerable variation of interpretation results, and the FORCE Project has established a subgroup ("Interpretation of the Location Referencing Rules") to resolve these problems.

7.5.6 Data Formats of the TMC Feature

Coding of TMC makes use of the ODA concept. Two coding protocols exist: ALERT-C, which will generally be used for the Pan-European service (PES) and which is described in ENV 12313-1, and ALERT-Plus with ALERT-C, which will be used in France and of which, in 1997, only a draft specification existed (prENV 278/4.1.1/0004).

The PES will operate in two different ODAs:

- The first ODA, dealing with ALERT-C only;
- The second ODA, allowing operation of a mixed ALERT-Plus with ALERT-C service described in the draft specification prENV 278/4/1.1.1/0004 or updates of it.

For this reason, the ALERT-C structure using type 8A groups is the same for both types of protocol—both applications have the same application group type 8A but a different application identifier code (AID). Table 7.3 illustrates the situation.

Figure 7.5 shows the format of type 8A groups when used for RDS-TMC. This group carries the TMC messages, service provider name, and tuning information.

The coding of ALERT-C messages is shown in Figure 7.6.

Table 7.3
Parameters for RDS-TMC Coding

Parameter	ALERT-C	ALERT-C With ALERT-Plus
ODA-AID in type 3A groups	CD46 hex	4B02 hex
Transmission modes	Basic or time windowing	Basic
Service	PES	ALERT-Plus and PES

When T (X4) = 0, the message data fields are either carried within single (X3 = 1) or multiple groups (X3 = 0), as shown in Table 7.4.

When T (X4) = 1, the 8A groups carry tuning information (variants 6 to 9) or the service provider name (variants 4 and 5), as shown in Figure 7.7. The service provider name is comparable to the PS name and it also has eight alphanumeric characters. To enable the TMC equipment to switch to another transmitter when the signal becomes weak, the tuning information indicates other networks (ON) that carry the same or another RDS-TMC service for a controlled handover.

Table 7.4
Bit Allocation for TMC Message Data

Data Field	Options	MSB	LSB
T(1 bit) :0	2	X4	—
F: Single/multiple group message identifier (1 bit)	2	X3	—
X3 = 1 single group			
X3 = 0 multiple groups			
DP: Duration and persistance (3 bits) for X3 = 1	8	X2	X0
CI: Continuity index (3 bits) for X3 = 0			
D: Diversion (1 bit) for X3 = 1 (single group)	2	Y15	—
Set to 1 for X3 = 0 (multiple groups)			
+/−: Direction (1 bit)	2	Y14	—
Extent (3 bits)	8	Y13	Y11
Event (11 bits)	2,048	Y10	Y0
Location (16 bits)	65,536	Z15	Z0

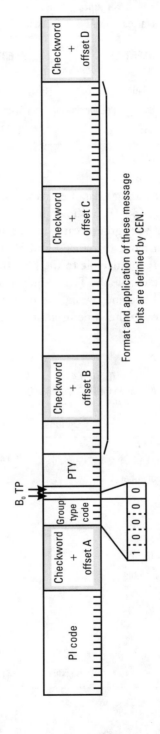

Figure 7.5 Type 8A group Traffic Message Channels. (*Source:* EBU.)

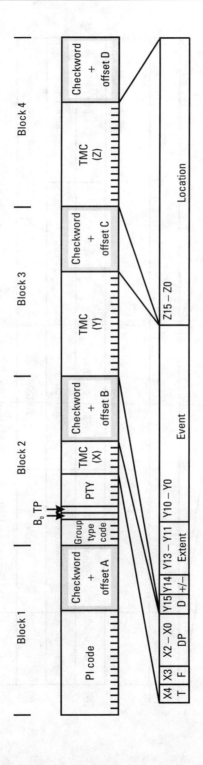

Figure 7.6 Coding of ALERT-C single group messages. (*Source:* EBU.)

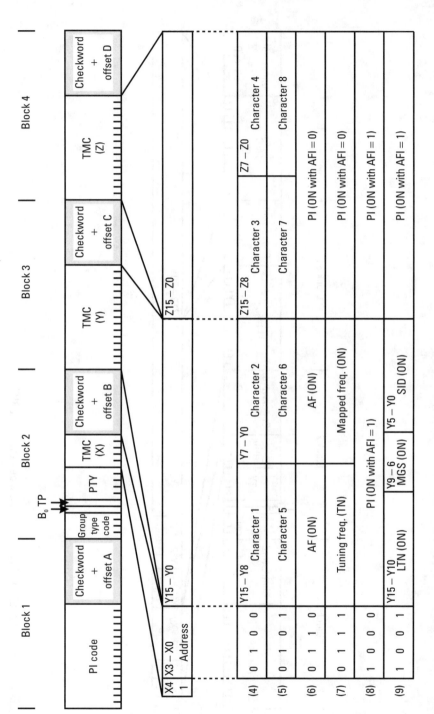

Figure 7.7 Data format for tuning information/service provider name. (*Source:* EBU.)

Variants 0–3 and 10–15 will be used for the ALERT-Plus protocol and are not further explained because of the still provisional nature of that specification.

The system information is coded in block 3 of group type 3A, as is shown in Figure 7.8.

The parameters used for system information are detailed in Table 7.5.

7.5.7 Principles of RDS-TMC Event Coding

Coding of road transport events for transmission in RDS-TMC is described in the CEN prestandard prENV 12313-2: May 1997. This specification contains the event list in the form of a database. The maximum number of events that can be coded is 2,048, but in that specification only 1,375 events have so far been coded.

Table 7.5
Parameters Used for System Information (Figure 7.8: Type 3A Group/Block 3)

Variant Code 00:			
LTN		6 bit	Location table number
AFI		1 bit	Alternative frequency indicator
M		1 bit	Transmission mode indicator
MGS:			Message Geographical Scope
	I	1 bit	International (EUROAD)
	N	1 bit	National
	R	1 bit	Regional
	U	1 bit	Urban
Variant Code 01:			
SID		6 bit	Service IDentifier
Transmission Mode Parameters:			
	G	2 bit	Gap parameter
	T_a	2 bit	Activity time
	T_w	2 bit	Window time
	T_d	2 bit	Delay time

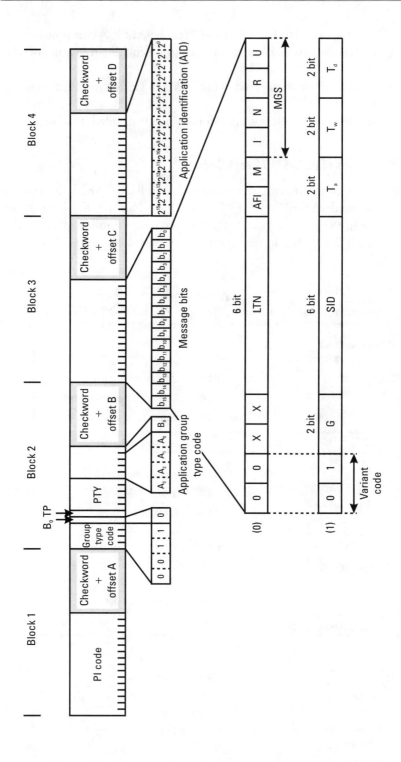

Figure 7.8 Coding of system information in group type 3A. (*Source:* EBU.)

To describe an event, we want to use an example: A passenger car had an accident and when the car parked at the roadside waiting for assistance to arrive, two dogs jumped out and ran on the motorway. With the ongoing traffic, the owners were not able to catch them and bring them back into the car. Passengers from other cars passing by noticed the hazard and informed the traffic service of the automobile association of the incident. The automobile association forwarded the message to the regional broadcaster, who sent out a message about this event using the appropriate location code. The message sent has the code 922, meaning "animals on roadway." The coding system does not permit specifying that the animals were dogs, or what kind of dogs. The same code is used for one or several animals and independent of the kind of animal—mouse, cat, dog, crocodile or elephant.

Because there are 1,375 possible codes, the database has been structured into 31 categories (also called update classes) with the view of assisting the message generation operator in quickly finding the right code—in this case, 922. This message belongs to category 13: Dangerous situations.

Table 7.6 contains the list of event categories used and also how many events are foreseen in each of these categories. One can see from the list that the editing software used by the operator for TMC coding can then be hierarchically structured along the line of these categories.

7.5.8 Principles of RDS-TMC Location Reference Coding

Most messages provide information about a location (e.g., a stretch of road, an intersection, or a region) and they refer to it by using a location reference. This is an identifier that can be interpreted without ambiguity by the receiving system. In RDS-TMC, locations are predefined and precoded, and the codes are stored in location code tables. The maximum number of codes in one table is determined by the field length for location codes in RDS-TMC; that is, 16 bits, which corresponds to 65,536 possible codes. The disadvantage of these tables is that they need to be created and maintained as was explained above, and the receiving system must use exactly the same table as the one used for encoding of the message. Otherwise, the message is not receivable.

The rules for location reference coding are specified in CEN prestandard prENV 278/7/3/0005: May 1997. These rules apply to ALERT-C messages only. ALERT-Plus uses a different system. Because of the provisional nature of ALERT-Plus in 1997, no further reference is made to that particular coding method, and all details that follow are extracted from the prestandard and apply thus to ALERT-C messages only.

Predefined locations are referenced by their location code, which is the tabular address of a number of prestored location details. Each table of stored

Table 7.6
List of Event Categories (Update Class) Used to Structure the Event Database and Number of Event Codes Falling Into Each of These Classes

Update Class	Events
1. Level of service	418
2. Expected level of service	58
3. Accidents	28
4. Incidents	14
5. Closures and lane restrictions	123
6. Carriageway restrictions	19
7. Exit restrictions	12
8. Entry restrictions	6
9. Traffic restrictions	56
10. Carpool information	11
11. Roadworks	30
12. Obstruction hazards	70
13. Dangerous situations	17
14. Road conditions	54
15. Temperatures	8
16. Precipitation and visibility	58
17. Wind and air quality	13
18. Activities	41
19. Security alerts	13
20. Delays	142
21. Cancellations	31
22. Travel time information	9
23. Dangerous vehicles	15
24. Exceptional loads / vehicles	20
25. Traffic equipment status	30
26. Size and weight limits	11
27. Parking restrictions	5
28. Parking	29
29. Reference to audio broadcasts	8
30. Service messages	22
31. Special messages	4
Number of events (max. 2048)	*1375*

locations must be given a unique location table number by one unique agency in each country or state. A country code (note: RDS-TMC uses the RDS country codes given in Appendix G) identifies the agency responsible for location reference coding and which one defined the location table and its number.

Many location references extend through several adjacent areas or road sections. The concept of primary and secondary locations is then used to indicate the extremities of the affected sections without having to list all the intervening places. For example, if an accident occurs at km 14.2 on the E15 (A26 road in France) and the resulting queue extends back to km 10.9, the situation location can be defined as E15, "km 14.2–10.9," where km 14.2 is defined as the primary location and km 10.9 is the secondary location. The primary location is taken to be where the cause of the problem can be found, whenever a cause can be pinpointed geographically. However, both primary and secondary locations will lie on the same road.

For the primary location, the location reference is the nearest downstream location in the direction of travel. The secondary location is indicated in terms of extent; that is, the number of steps back along the road through other predefined locations. Alternatively, a distance marker may be used.

All location codes belong to a unique location table. Within any particular location code table, each location has one unique number in the range 1–63,487. The other 2,048 numbers are reserved for EUROAD, an agreed upon concept used for coding messages to international travellers on the Trans European Road Network (TERN).

RDS-TMC uses a hierarchical structure of predefined locations. A system of pointers provides upward references to higher level locations containing the specified location. For example, Kent would have an upward area reference to southeast England, which will be upwards referenced to the United Kingdom, then the British Isles, then Europe.

Junction 25 on the M1 motorway in the United Kingdom would have a section of route referenced to a motorway segment (e.g., Leicester–Sheffield). This segment will then be referenced upwards to the whole road (i.e., the motorway M1).

Also, Junction 25 on a motorway may be offset to Junction 26 in the positive direction, and to Junction 24 in the negative direction.

In many cases, events affecting road traffic cover a number of locations, such as where an accidents results in long tailbacks. The ALERT-C protocol defines such occurrences by addressing the location of the accident as the primary location, then identifying the end of the tailback by using the direction and extent fields. These fields consist of four bits in total: one direction bit and three extent bits. The direction bit indicates the queue growth and not the direction of traffic flow. The extent bits identify the number of locations along the road that are affected by the problem, with a maximum of eight (primary location and seven related locations). An extent of 1 would identify the secondary location (the end of the event's extent) as being the next location along the same road from the primary location. An extent of 3 would force the receiver to search the database for the third location along the same road from the primary location as is illustrated in the example given in Table 7.7.

Three additional optional columns exist. These are the motorway exit number and the deviation reference road number in positive and negative directions to give diversion advice.

Table 7.7
Coding Example for Location Referencing

Location Code	Type	Road Number	Name 1	Name 2	Ref A	Ref L	Neg. Offset	Pos. Offset
949	L3	E1	X-town	Y-town	2009	—	948	950
2009	A6.2	—	Greater neighbourhood	—	1	—	—	—
4420	P3.2	E1	Bridge	—	2009	949	4456	4423
4423	P1.3	E1	Place A	N207	2009	949	4420	4459
4459	P3.3	E1	Parking	—	2009	949	4423	4460
4460	P1.3	E1	Place B	—	2009	949	4459	4461

Location code: Identifier in the range 0 to 65,536 describing a stretch of road, an intersection or a region;

Type: Defines the kind of location in three main categories and a number of sub-categories: Areas (A), Linear locations (L), and Points (P). In the example given above, L3 means a road link, A6.2 a metropolitan area, P3.2 a bridge, P1.3 a motorway junction, and P3.3 a service area;

Road number: The road reference number;

Name 1: Name of the primary location that will be presented by the receiver;

Name 2: Name of a secondary location—required to describe a road segment;

Ref A: Pointer to the area to which the location belongs;

Ref L: Pointer to the stretch of road to which the location belongs;

Negative offset: Pointer to the previous area or location (e.g., on the same stretch of road);

Positive offset: Pointer to the next area or location (e.g., on the same stretch of road).

7.5.9 Example for Constructing an RDS-TMC Message

The message:

> Motorway A9 Munich-Nuremberg, direction Nuremberg, stationary traffic between exit Pfaffenhofen and motorway interchange Holledau. Deviation recommended via the U31 from exit Pfaffenhofen,

requires the following message elements, which are given here as decimal numbers and which will then be binary coded:

- *Duration:* 0 for no specific duration specified;
- *Diversion:* 1 for deviation being recommended;
- *Direction of event:* 1 for negative;
- *Extent of event:* 2 for two locations backward in the location code list;
- *Event:* 101 for stationary traffic;
- *Location code:* 12735 for motorway interchange Holledau.

The coded type 8A group for this example is shown in Figure 7.9.

The corresponding part of the location code list is given in Table 7.8. For simplification of understanding, all numbers are given as decimal numbers, but in reality these are coded as binary numbers.

7.5.10 RDS Encoders and the EBU/UECP

After editing the RDS-TMC messages in the broadcasting house, they are sent to the transmitter or the transmitter network for insertion into the RDS data stream. This task is achieved with the help of RDS encoders that receive the messages in a specified format, as shown in Figure 7.10.

RDS encoders certainly need software adaptation to allow RDS-TMC to be implemented as it is now specified. One of the protocols of interest from the European point of view is the EBU Universal Encoder Communication Protocol (UECP), which required an urgent enhancement to include all the newly defined possibilities specified.

The EBU document SPB 490 (version 5.1) is an upgrade to achieve the same enhanced functionality (especially for ODA and TMC) as offered by EN 500067:1998 and the most recent RDS-TMC specifications. This work was started within the RDS Forum in 1996 and completed during 1997.

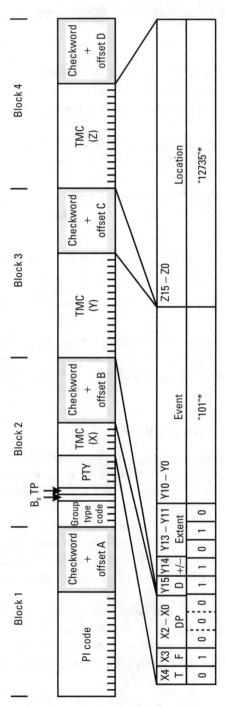

* Use binary coded decimal number instead

Figure 7.9 Coding of the example in type 8A group. (*Source:* EBU.)

Table 7.8
Extract From the German Location Code List (The bolded terms are those used in the example)

Location Code	Type	Road No	Sub Type	Name 1	Name 2	RefA	RefL	Off+	Off–	Exit No	U+	U–
256	BL			Schleswig-Holstein	1							
257	BL			Hamburg	1							
258	BL			Nieder-sachsen	1							
259	BL			Bremen	1							
260	BL			Nordrhein-Westfalen	1							
261	BL			Hessen	1							
262	BL			Rheinland-Pfalz	1							
263	BL			Baden-Württemberg	1							
264	**BL**			**Bayern**	**1**							
265	BL			Saarland	1							
266	BL			Berlin	1							
267	BL			Brandenburg	1							
268	BL			Mecklenburg-Vorpommern	1							
269	BL			Sachsen	1							
270	BL			Sachsen-Anhalt	1							
271	BL			Thüringen	1							
—	—	—	—	—	—	—	—	—	—	—	—	—
—	—	—	—	—	—	—	—	—	—	—	—	—
292				Münster	260							
293				Niederbayern	264							
294				**Oberbayern**	**264**							
295				Oberfranken	264							
296				Oberpfalz	264							
297				Rheinhessen-Pfalz	262							

Table 7.8 (continued)

Location Code	Type	Road No	Sub Type	Name 1	Name 2	RefA	RefL	Off+	Off−	Exit No	U+	U−
299				Schwaben		264						
—	—	—	—		—	—	—	—	—	—	—	—
—	—	—	—		—	—	—	—	—	—	—	—
602				**Pfaffenhofen a. d. Ilm**		294						
603				Pforzheim		284						
—	—	—	—		—	—	—	—	—	—	—	—
—	—	—	—		—	—	—	—	—	—	—	—
7219	**SEGME**	**A9**	**LINE**	**München**	**Nürn-berg**	**264**	**264**	**7220**				
7220	SEGME	A9	LINE	Nürnberg	Halle/ Leip.	264	270	7221	7219			
—	—	—	—		—	—	—	—	—	—	—	—
—	—	—	—		—	—	—	—	—	—	—	—
12725	POINT	A9	AS	München-Kieferngarten		549	7219	12726	12724	73		
12726	POINT	A9	AK	München-Nord		549	7219	12727	12725	72		
12727	POINT	A9	AS	Garching-Süd		549	7219	12728	12726	71	25	78
12728	POINT	A9	AS	Garching-Nord		549	7219	12729	12727	70	25a	76a
12729	POINT	A9	AS	Eching		418	7219	12730	12728	69	27	76
12730	POINT	A9	AK	Neufahrn		418	7219	12731	12729	68		74
12731	POINT	A9	TR	Fürholzen		418	7219	12732	12730			
12732	POINT	A9	AS	Allershausen		418	7219	12733	12731	67	29	74
12733	**POINT**	**A9**	**AS**	**Pfaffenhofen**		**602**	**7219**	**12734**	**12732**	**66**	**31**	**72**
12734	POINT	A9	TR	In der Holledau		602	7219	12735	12733			
12735	**POINT**	**A9**	**AD**	**Holledau**		**602**	**7219**	**12736**	**12734**	**65**	**33**	**70**
12736	POINT	A9	AS	Langenbruck		602	7219	12737	12735	64	35	68
12737	POINT	A9	AS	Manching		602	7219	12738	12736	63	37	66
12738	POINT	A9	AS	Ingolstadt-Süd		477	7219	12739	12737	62	39	64
12739	POINT	A9	AS	Ingolstadt		477	7219	12740	12738	61	41	62

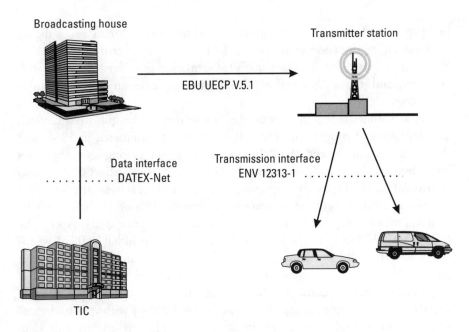

Figure 7.10 Routing of RDS-TMC messages from the TIC to the end users. (*Source:* EBU/IRT.)

The encoder model and protocol (see Chapter 11) provide a template specification upon which new products may be based and other existing encoder communication protocols may be enhanced. Thus, many existing devices could be adapted to meet the functionality required, and subsequently, a universal protocol may be implemented. However, encoders of the first RDS generation can generally only be upgraded with significant software modifications, and it is difficult, or even impossible, to achieve the full support of the present level (version 5.1) of the UECP protocol. For this reason, the introduction of RDS-TMC may also require substantial investments to be made on the side of the transmitter network to replace the older RDS encoders (e.g., about 800 for a large country like France). This is likely to lead to a further delay in the implementation of RDS-TMC.

7.5.11 RDS-TMC Receivers

7.5.11.1 European Industry Position

Major European car radio manufacturers have collaborated in the European Association of Consumer Electronic Manufacturers (EACEM) for many years to develop a European car radio technology meeting the objectives mentioned, but also with the objective to support car navigational systems using digital road maps in conjunction with updated TTI. These manufacturers require that European standards be used by the TTI information providers, the car manufacturers, and the authorities regulating the implementation of this new technology.

Car manufacturers in Europe also cooperated in the industrial research and development programme PROMETHEUS that originated in 1986, and that terminated at the end of 1994. This programme had long-term objectives going beyond the year 2000, especially aimed at the application of new (also electronic) technologies to improve a variety of things from the safety of car driving to all sorts of enhancements like improved vision (in bad weather), collision avoidance, vehicle communication, and navigational assistance. The RDS-TMC objectives described above have also been included in the development, and thus all car manufacturers in Europe have gained experience with RDS-TMC and are now prepared to support its implementation.

The expectation is that by the year 2000 it will be common practise that cars will be increasingly equipped with the new technology with various degrees of perfection, not only concerning the car radio but also navigational aids, electronic maps, onboard computers, and travel planning systems. Since the information to be distributed primarily concerns messages that need to be updated continuously—concerning millions of road users—it appears to be sensible to look at the use of broadcast technology for disseminating TTI to the user. The only relevant broadcast technology available now and harmonised on the European scale is RDS; hence, the expectation of using RDS-TMC as early and as widely as possible.

7.5.11.2 Receiver Requirements

Ideally, RDS receivers should have an interface with RDS data to permit devices with new services such as TMC or DGPS to be connected to existing RDS radios. Unfortunately, in the large majority of existing RDS receivers, such an interface does not exist, mainly because no agreement could be reached to standardise it—and an extension of existing RDS receivers to TMC functionality is therefore impossible.

New TMC receivers should have a double-tuner front end to permit the end user to freely choose a radio programme channel and then receive the TMC service in the background.

There should also be on all TMC receivers a standardised input for location code tables; that is, the smart card to be used should be standard or, alternatively, there should be an industry standard.

TMC receivers should implement message management; that is, deal with message presentation according to the corresponding degree of urgency and automatically delete old messages, making use of the time stamp derived from Clock Time (CT) transmitted in type 4A groups.

A filter for relevant messages will have to be implemented in TMC receivers, where relevant means relevant to the journey undertaken.

The message presentation will have to take road safety requirements into account. Speech output needs to be of a sufficiently high audio quality—comparable to FM radio—to assist the perception of traffic announcements without additional stress. Scrolling text on a display is generally considered as dangerous because of the distraction that it can create for the driver, although it will be easy and inexpensive to implement it in less expensive TMC receivers.

7.5.11.3 First-Generation RDS-TMC Receivers

7.5.11.3.1 Car Radios

The only RDS-TMC car radio available in 1997/98 with speech output is the Blaupunkt model Viking TMC 148, as shown in Figure 7.11.

This radio tries to offer a reasonably well-designed TMC output, but to keep the price of this product down between USD $400–$500/

Figure 7.11 Blaupunkt Viking TMC 148. (*Source:* Blaupunkt.)

DEM 700–900, retail), a number of technological constraints have negatively influenced the design. These are as follows:

1. Only a single tuner is used, which means for TMC that tuning to the TMC channel is obligatory.

2. The only available language is German, and other language versions were not announced at the initial launch date.

3. Speech output is limited to only 68 message types and only a few locations; other messages can only be presented on the eight-character display used by the car radio and also for the PS name.

4. The exit locations on motorways cannot be identified by name; exits are identified by numbers instead.

On the positive side, the radio can filter messages and select the desired motorways or national roads. This information has to be keyed in, which is a bit difficult to achieve for the driver. The speech output is of a reasonably high quality, since it uses recorded and not synthesised speech. The smart card for Germany contains location codes for all motorways and national roads, and also some major roads in urban areas (14,000 locations altogether stored on 512 Kb). The smart card is supplied with the receiver and is included in its price.

Given the technological limitations of this product, there will certainly be problems in marketing it on a European scale. Since the receiver was conceived only for the German market, it will probably be difficult to subsequently adapt it to the European market, which would be a necessity to achieve production in large quantities at a retail price that the end consumer will find attractive. It is thus possible that this product will remain a kind of demonstrator of the limited possibilities imposed by present state technology.

The limitations encountered with this product will most probably apply to the whole industry. It was confirmed in 1997 that it still was not possible to manufacture an RDS-TMC radio at a reasonably low price and meet the ideal requirements for such receivers that would correspond to the service possibilities offered by the RDS-TMC system that had been in development since 1984. This kind of problem was already anticipated some years before [12].

Philips Car Systems announced at IFA '97, also in Germany, their first RDS-TMC car radio, RC 579, shown in Figure 7.12. However, this radio cannot present TMC messages through speech; it generates corresponding icons that symbolise the traffic message content, which are sent to the CARIN 520 navigation system that inserts them on the map display. The CARIN navigational system uses a digital road map atlas stored on a CD-ROM that also contains the ALERT-C location codes. The announced retail price for the radio

Figure 7.12 Dynamic route navigation with RDS-TMC Philips RC 579 + CARIN 520. (*Source:* Philips.)

was approximately USD $500/DEM 800 and the CARIN navigation systems costs USD $3,000/DEM 4,800, in addition. The tuner has only one front-end, and, as in the case of the Blaupunkt radio, tuning to the RDS- TMC channel is compulsory if the dynamic route information is required by the driver.

7.5.11.3.2 Navigational Systems

A Volvo/Mitsubishi RTI product is one of the first navigational systems developed independent of the car radio that has dynamic updating via RDS-TMC (see Figure 7.13). This system is factory fitted as an option for Volvo cars and works like the Philips system described above. However, the receiver has two tuners, one for the car radio and another for RDS-TMC. This system was released in 1997 and its retail price is approximately USD $3,600/SEK 28,000. Different CD-ROMs are necessary for different European countries in which the system is marketed. Apart from the traffic information presented through icons, the system also displays some tourist information (hotels, petrol stations, Volvo workshops, etc.), which is also stored on the CD-ROM.

Figure 7.13 Volvo / Mitsubishi RTI navigation system with RDS-TMC. (*Source:* Volvo.)

7.5.11.3.3 ALERT-Plus Receivers

During the years 1995–1997, there was a flurry of activity that was limited to France and that was aimed at developing the RDS-TMC ALERT-Plus protocol, which is status message oriented, requiring continuous updating. This has confused the situation about the availability in France of a Pan-European RDS-TMC service, but it now appears the plan is to operate it there in parallel with ALERT-C, which is event message oriented. ALERT-Plus is a proposal originating from the DRIVE II period, and was promoted in France to provide a TTI service carrying status information, such as car park capacity and anticipated route travel-time information. This nonevent-oriented service appears to require significantly more data capacity, and it has been suggested that it is more suitable for DARC or digital radio (DAB). But in France, particularly in Paris, several former CARMINAT partners (mainly Télédiffusion de France (TDF), Sagem, Renault, and the recently created consortium Médiamobile) appear to believe that the RDS capacity will be sufficient, at least for starting a service in the Paris region. In other locations where broadcasters use much more RDS capacity for their own programme-related needs, especially when RadioText is implemented, it is unlikely that such a service can be operated using RDS.

In early 1998, Médiamobile used approximately five RDS groups per second on each of three frequencies with high-power transmitters radiating from the top of the Eiffel tower. In this way, full coverage of the Paris metropolitan area is achieved. One of these transmitters carries travel time status data, and the other two carry the data about traffic flow/congestion for all the major roads in the Paris metropolitan area. That information is divided between two transmitters, one giving the information for the centre of Paris and the other for the outskirts (see Figure 7.14). The data transmitted by Médiamobile is prepared by combining the information from two sources. Firstly, data is collected by the city of Paris using magnetic loop detectors, which are installed every 500 to 750m on 2,000 km of all major roads in Paris. Secondly, floating car data is collected by one taxi operator that coordinates some 2,000 taxis.

The first ALERT-Plus receivers were presented by Sagem in 1997, as shown in Figure 7.15. Médiamobile started with these receivers: one handheld terminal called TM 2000 Visionaute, and the other available as a factory-fitted

Figure 7.14 The traffic information centre from the city of Paris. (*Source:* EBU.)

Figure 7.15 The Sagem/Médiamobile Visionaute is designed for a pay-service in Paris using the ALERT-Plus with the ALERT-C protocol. (*Source:* EBU.)

model for Renault's new car model Megane Scenic, a TTI pay service, as illustrated in Figure 7.16. This service started at the end of 1997 in the Paris metropolitan area and is planned to become available later in other major French cities. The price of the portable receiver is approximately USD $520/FRF 3,000, retail, and the integrated one was announced to cost approximately USD $1,400/FRF 8,000. The service subscription costs approximately USD $20/FRF 120 per month.

During 1997, the French car manufacturer PSA Peugot/Citroën and the French service provider CGEA/Eurolum presented their own ALERT-Plus system product, INF-FLUX. It has a modified protocol that they claim is less demanding on RDS capacity and achieves the same objectives as the Médiamobile service, with which it wants to compete.

7.6 Alternative Technologies

7.6.1 GSM

Studies to use the Global System for Mobile communications (GSM) for the provision of two-way communication between information centres and

Figure 7.16 Renault's Megane Scenic receiver for the Médiamobile pay service. (*Source:* EBU.)

computers onboard equipped vehicles started in the European Commission's DRIVE I (1989-1991) project: System of Cellular Radio for Traffic Efficiency and Safety (SOCRATES) [4].

GSM includes the possibility for data communication such as the short message services (SMS) channel, which can be used for traffic message services, point-to-point and point-to-multipoint broadcast (cell broadcast), route guidance, and emergency calls. GSM also provides a full data call functionality that permits implementation of pre- and on-trip travel planning and also dynamic route guidance in a navigational system. The widely spread GSM technology is thus suitable for a large variety of ITS services, either using the bidirectional communication link or the inherent data broadcast features.

The switched GSM data service permits a user data rate of up to 9.6 kbps. SMS permits the transmission of short datagrams with a length of 140 bytes. The use of concatenated SMS permits transmission of even longer messages [13].

During the years 1996–1997, a new structuring of the SMS message service was introduced (user data header concept). This concept allows easy and efficient application protocol addressing by port numbers. Furthermore, this concepts allows fragmentation and reassembling of long messages, concatenating up to 255 SMS messages into one message. One example of a transport layer protocol that makes use of this data header concept is the narrowband socket (NBS) approach.

New solutions for higher data rates on GSM were already standardised in 1997. These standards included a high-speed circuit-switched data (HSCSD) service and a general packet radio service (GPRS).

Integration of Internet and GSM technology has been studied extensively in the EC-funded PROMISE project. This project is aimed at the development of a traffic and travel information service employing a new type of user terminal, the Nokia 9000 Communicator, using the data communication capabilities of the cellular GSM network [14].

Finally, the Universal Mobile Telecommunication System (UMTS) will represent the third generation of wireless mobile voice and data communication. This will probably attain a user data rate of 2 Mbps. The wide availability of GSM makes it a good starting platform for the introduction of UMTS services [15].

The European Telecommunication Standards Institute (ETSI) standardisation work provides for a phased approach of evolutional enhancements to the GSM technology with the view of ensuring a long lifetime for GSM products using earlier versions of the standards.

With particular regard to ITS services, in early 1997, ERTICO—which is a European ITS technology interest group formed by manufacturers, service providers, and road authorities—created an ITS services committee whose members were major European suppliers and service providers. The aim was to define a strategy for supporting ITS services via GSM [16]. In the report of this committee, two important developments were identified that enable GSM to carry ITS services efficiently:

1. The first development is the emerging industry standard (in 1997, forwarded to CEN TC 278) for telematics services, called Global Automotive Telematics Standard (GATS), which was jointly developed by two major European telematics service providers: Mannesmann Autocom and Tegaron (a company created jointly by Deutsche Telekom and Daimler Benz Interservices) [17]. The emerging standard was already implemented in systems offered in Germany 1997–1998. GATS makes it possible to offer a group of commercial ITS services on GSM. For data transmission, a modular message structure has already been specified on application-oriented layers of the air interface between the service centre and the mobile station. It is subdivided into a transport protocol, conditional access, a security layer, and application protocols containing service-related data.

2. The second development is a proposal on generic wireless information access, called Wireless Application Protocol (WAP) [18]. WAP can coexist with GATS and extends the functionality towards Internet-like information access, permitting the use of channels with a limited data rate. It thus provides a new access medium to ITS services on GSM. The first version of the standard was released in September 1997. A WAP Forum will be created to manage the further development and implementation of this standard.

The ERTICO committee identified a list of ITS services that can be provided via GSM. These are as follows:

1. Emergency and breakdown services/assistance and security services;
2. Traffic information;
3. Floating car data, based on individual measurements of telematics onboard units within individual cars. This can be seen as traffic monitoring which can be implemented at reasonable infrastructure cost;
4. Route guidance;
5. Fleet management applications;
6. Information services to provide the driver with individual information;
7. Vehicle theft detection and recovery;
8. Vehicle remote diagnostics.

GSM coverage is generally continuous and covers most of Europe. It is designed to operate in an urban and interurban environment. GSM permits easy implementation of commercial ITS services since the system permits roaming, handover, and SIM-based authentication. Examples of GSM integrated car receivers are shown in Figures 7.17 and 7.18.

Figure 7.17 Blaupunkt integrates in the GEMINI a car radio, GSM and GPS. (*Source:* Blaupunkt.)

Figure 7.18 In 1997, Mannesmann-Autocom bought Philips Car Systems, now Mannesmann-VDO Car Communications, to market telematics car radios combined with telematic processors using GSM and GPS technology. (*Source:* Philips.)

In 1997, during the 4th World ITS Congress in Germany, a number of new German ITS services using GSM had already started regular operation or were announced [19]. These are as follows:

- *OnStar:* Buyers of new Opel cars can order an info set USD $800 (DEM 1,500) consisting of a mobile phone, hand-free equipment for fixed car installation, and a global positioning system (GPS) receiver. By the push of a button, the driver is connected with the German automobile club, ADAC, service centre. The vehicle position is automatically transmitted, and the driver can then be guided online to the desired destination. The monthly fee is approximately USD $16/DEM 30.00 and each connected minute costs approximately USD $1.30/DEM 2.50.

- *Mannesmann-Autocom 22666 or Passo:* Provides traffic information over the German D2 network through voice dialogue with a computer system. This service costs an extra fee of approximately USD $0.83/DEM 1.56 per minute. A special receiver with GPS (USD $800/DEM 1,500) will be launched, and via the VDO "mobility assistant" MobiMax, destination direction and distance will also be communicated.

- *Tegaron:* A new company formed jointly by Daimler Benz Interservices and Deutsche Telekom in cooperation initially with Blaupunkt and Mercedes Benz. In January 1998, automated emergency call and theft detection services started. The special equipment needed, which includes a GPS navigational unit, is offered as an option on certain Mercedes cars. The services offered are available as a pay service. Dynamic route guidance and individual traffic information are additional service options.

- *ADAC:* The German Automobile Club offers a service for its members providing traffic information and route assistance. The possibility of offering an attractive, reasonably priced receiver is being studied.

The scenarios for all these new TTI services and their inter-relationships are shown diagrammatically in Figure 7.19.

7.6.2 DARC (also previously known as SWIFT)

The System for Wireless Infotainment Forwarding and Teledistribution (SWIFT) was developed in the Eureka 1197 project. The aim was to develop a multiapplication data system using the FM radio network for mobile and portable receivers. In 1997, the SWIFT Forum decided to only use the name DAta Radio Channel (DARC) instead of SWIFT to avoid confusion with an international organisation using precisely that name.

The DARC system is actually seen as a system that has the potential to provide an interim data service during the transition period from RDS to DAB data broadcasting. DARC technology is inexpensive, and full advantage can be taken of the existing FM coverage. However, when FM networks are very

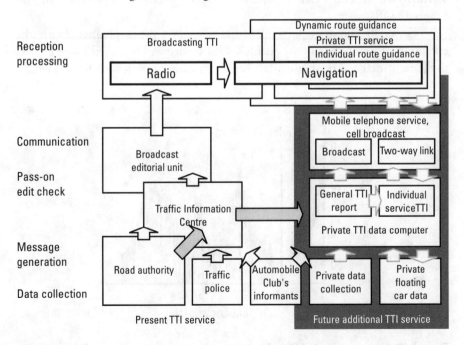

Figure 7.19 Upcoming new scenario for TTI service provision. (*Source:* EBU.)

dense, DARC may create adjacent channel interference, a matter that is still under investigation, which is one of the reasons why the system is not yet very popular among the broadcasters operating large FM transmitter networks.

The SWIFT project adopted the physical and data link layer of the DARC system, developed by the Japanese broadcaster NHK and the Swedish transmission operator Teracom [20, 21].

DARC's modulation type is level-controlled minimum shift keying (LMSK). The −40-dB bandwidth is 44 kHz centred around 76 kHz, and the overall bit rate is 16 Kbps. Figure 7.20 shows the baseband spectrum of an FM sound broadcast multiplexed with RDS and DARC.

The subcarrier level is dependent on the level of the stereo difference signal. Because the amount of multipath interference has been shown to depend on the modulation level of the stereo signal, a high injection level of the subcarrier is required to provide good reception quality when the stereo sound modulation is high. Typically, a maximum injection level of ±7.5 kHz is used (RDS typically uses only ±2.0 kHz).

In 1997, the following European countries and transmission operators considered the introduction of regular data services using DARC as standardised by ETSI from 1998 onwards: Sweden (Teracom), Norway (Norkring), Germany (Deutsche Telekom), Austria (ORF), Netherlands (Nozema), and France (TDF).

Generally, no transport telematics services are yet planned, with the exception of France where the ALERT-Plus service will be tested using DARC to take advantage of the higher data capacity in comparison with RDS-TMC.

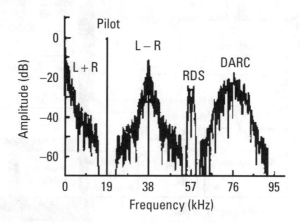

Figure 7.20 Multiplex spectrum of FM stereo sound broadcasting with RDS and DARC. (*Source:* EBU.)

In Japan, the DARC system is used in conjunction with radio and infrared beacons for the operation of the Vehicle Information and Communication System (VICS) service, which has provided traffic information to in-vehicle units since 1996 in three large metropolitan areas (Tokyo, Aichi, and Osaka) [22, 23].

Three different service levels are in operation:

- *Level 3:* Colour map display giving traffic density information;
- *Level 2:* Simple graphic display depicting traffic information;
- *Level 1:* Text information essentially giving travel time.

The system is very widely supported by some 27 consumer electronics manufacturers, offering 74 types of onboard VICS products in 1997. In 1996, many of the navigational systems sold in Japan (800,000 units) could use VICS for dynamic updating. There are no plans yet in Europe to use VICS, since RDS-TMC will give similar functionality.

7.6.3 DAB Delivery of TMC Messages

While Sweden expects to cease FM radio broadcasts in favour of digital broadcasting within 10 years, few other countries expect to make the change that quickly, even though most European countries are now launching digital radio services (as are Canada, Australia, and South America).

Digital audio broadcasting (DAB) promises broadcast audio quality close to that of compact disc and includes several data stream options that can be used for additional data services such as TMC. The message capacity of DAB is potentially as much as 1,000 times greater than RDS-TMC. Although the standards for DAB have been adequately developed for audio broadcasting, they have not yet been fully developed for additional data. In 1997, various field trials were undertaken to evaluate the use of road maps delivered by DAB using the multi-media object-transfer (MOT) protocol for presentation on LCD displays.

DAB-TMC, in particular, needs more work, and the European Broadcasting Union is coordinating that work in the B/TPEG project, which is supported by a wide range of consumer electronics manufacturers, navigation systems companies, and service providers.

It is interesting to note that the European Broadcasting Union Statement D81-1996 recommends that broadcasters should develop their infrastructures for RDS-TMC with the expectation that the infrastructure can be used for DAB later. Thus, it will be important to develop systems especially for TTI message generation that provide output to one or more transmission delivery possibilities and to include a smooth path for migration to DAB.

In 1997, 100 million people in Europe could already receive pilot DAB transmissions if they had the appropriate DAB receivers—but the receivers are so expensive that the market has not generally accepted them, although they have been available for a number of years. Nevertheless, the development of DAB transmission networks will take a significant amount of time to achieve coverage that matches that of FM radio. For example, in the United Kingdom, the BBC plans to invest GBP 10 million/USD $16 million to achieve coverage for 60% of the population by the end of 1998. It could easily take another five years or more to reach 95% coverage, and even that is not as good as the present FM coverage.

7.7 The Longer Term Future of TMC

RDS-TMC was given a huge boost in 1997 when the European Union (EU) raised the issue of implementation of services to a high political level, giving very strong support to those involved in the final development phase of RDS-TMC. The first services were being announced in the same year, and in Germany, services were beginning to reach completion and were ready for consumer use. While it is difficult to predict how long TMC will be delivered by RDS, it can be assumed that this high-level EU concern will result in widespread adoption of the technology, and thus it can be expected that RDS-TMC will be available for many years to come. The possible development of TTI services delivery is shown in Figure 7.21.

However, the increased availability of the TTI data gathered for these new TMC services suggests that there will be a surfeit of messages, especially in the metropolitan areas, which RDS-TMC will not be able to handle due to the limited bandwidth available in RDS. Furthermore, there is now the possibility of using many different delivery mechanisms, such as DAB, DARC (also known as SWIFT), digital video broadcasting (DVB), GSM, Internet, and teletext in analogue TV. This new delivery environment expansion was recognised by the EBU in Statement D81-1996 [24], which recommends that broadcasters should develop their infrastructures for both RDS and DAB. But the EBU is taking an even more progressive step, having authorised a new development project known as B/TPEG, which is briefed to develop the Transport Programme Experts Group (TPEG) Protocol [25, 26]. In the short to medium term, the public broadcasters of the EBU support the B/TPEG activity to achieve a suitable protocol for DAB, and expect many other useful proposals to come from the B/TPEG project.

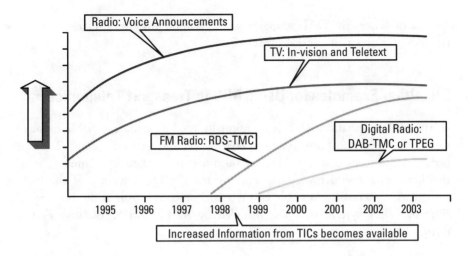

Radio: Voice Announcements

TV: In-vision and Teletext

FM Radio: RDS-TMC

Digital Radio: DAB-TMC or TPEG

1995 1996 1997 1998 1999 2000 2001 2002 2003

Increased Information from TICs becomes available

Figure 7.21 Possible development of broadcast traffic and travel information. (*Source:* EBU.)

The TPEG protocol will, in particular, enable broadcasters to develop their TTI services in nondelivery-specific databases, then undertake just one editorial activity and offer these TTI services via one or more delivery technologies, safe in the knowledge that no on-air conflicts of information will result. Use of a single TPEG protocol by all EBU members is expected to facilitate reduced production costs, which will be important for those broadcasters who continue to be TMC service providers as part of their public service portfolio.

A key idea of TPEG is to allow the generation of messages without the need for a location database, eliminating the need to carry location codes in a receiver. This will be designed so that various delivery technologies can carry the same message without further compression. However, low-bandwidth delivery technologies such as RDS could perhaps also be fed with TPEG messages via an adaptation layer, which will compress the information into messages that can be accommodated in the lower bandwidth.

The B/TPEG Plenary Group comprises a wide range of partners and includes EBU members, consumer electronics manufacturers, navigation systems companies, service providers, and transmission operators. It will first evaluate other related protocols and seek to use the best elements of all those known to be available. It will then produce a specification for the TPEG data stream as a draft standard, and produce a parallel guidelines document. Additionally, a TPEG messaging specification for event- and status-oriented road messages and a TPEG service identification and access control specification will be produced by late 1998, together with a parallel guidelines document. Then, the DAB adaptation layer will be produced in close collaboration with Eureka 147 and WorldDAB.

7.8 Other Examples for Using RDS in Transport Telematics

In transport telematics, RDS is occasionally used to update variable message signs, car park status information boards, (as shown in Figure 7.22) and boards at bus stops that signal bus arrival times. In any new implementation of this kind, one would nowadays generally use the ODA feature of the RDS system. An interesting example of such an implementation is the bus arrival time signalling system that was installed by AZTEC in Strasbourg, France during 1997.

Figure 7.22 Car park status information in Strasbourg updated via RDS-ODA. (*Source:* Info Telecom.)

7.9 Conclusions

As has been shown, RDS-TMC has a number of unresolved problems that explain to a certain degree the difficulties encountered in its general introduction. As it is generally known, each technology has an optimal time window where it can succeed; and after some time, other and better performing technologies will take over. When these appear on the horizon, it is false to conclude that the knowledge about the better technology is already the end of the so far believed good solution.

In consumer electronics, it takes years for the market to succeed with a new technology. Although receiver prices and profit margins for the industry are relatively low, competition and quality of the products is usually high and the lifetime of the equipment is long.

Success is only possible when well-coordinated standards exist, the technology to be used is mature, there is full consensus of all parties involved in service provision, and the whole industry is ready to go into mass production. In addition, there must be a high demand by the end users for a product that is attractive, safe for the driver, and really useful on the road. As we have seen, RDS-TMC remains surrounded by many open questions. Final answers about success or failure are not yet possible. The infrastructures required and built to collect traffic and traveller information with the view of RDS-TMC service provision will be needed anyway, regardless of the advent of new alternative technologies. Without any doubt, they represent useful long-term investments that are absolutely necessary for high-quality service provision over large areas such as the European Union or the United States.

For RDS-TMC to succeed, we need stable and widely agreed upon standards for the consumer electronics industry. The way their development has been coordinated by the European Commission has not, unfortunately, yet led to the success expected. What lesson can now be learned from 15 years of development effort put into this technological concept? One thing is certain: Those who have participated in this experience can now contribute to the development of better solutions that will use technologies such as DAB, DARC, and GSM.

References

[1]   European Commission, Community Strategy and Framework for the Deployment of Road Transport Telematics in Europe and Proposals for Initial Action, Brussels, 20.05.97 COM(97) 223 final.

[2]   Resolution on Standards for Traffic Messages broadcast using the Radio Data System
      Traffic Message Channel, The Council of Ministers of ECMT, Antalya, May 22-23,
      1991.

[3]   Council Resolution of the Sept. 28, 1995 European Commission, Official Journal
      95/C 264.

[4]   Catling, I., *Advanced Technology for Road Transport: IVHS and ATT*, Norwood, MA:
      Artech House, 1994.

[5]   Bücken, R., "Verkehrsfunk auf Abruf," *Funkschau* No. 20, 1997, pp. 70–73.

[6]   VDA: Tatsachen und Zahlen aus der Kraftverkehswirtschaft-59. Folge, Frankfurt/M.,
      1995.

[7]   ERTICO, TELTEN Final Report Vol. 1, European Commission DG VII, Brussels, 1994.

[8]   Ungerer, D., July 1997, not yet published.

[9]   European Conference of Ministers of Transport (ECMT) document, New Information
      technologies in the Road Transport Sector-Policy Issues, Ergonomics and Safety.

[10] British Standards Institute, DD 235: Draft for Development-Guide to In-vehicle Infor-
      mation Systems, 1996.

[11] Agreement on Guidelines for the Design and Installation of Information and Communi-
      cations Systems in Motor Vehicles, German Economic Forum on Telematics in
      Transport, Nov. 1996.

[12] Kopitz, D., "Development of the Traffic Message Channel in RDS," EBU *Technical
      Review*, No. 245, Feb. 1991, pp. 29–32.

[13] Redl, S. M., M. K. Weber, and M. W. Oliphant, *An Introduction to GSM*, Norwood, MA:
      Artech House, 1995.

[14] Geen. P., "Standardisation of Traveller and Traffic Information via Cellular Networks,"
      4th Word Congress on Intelligent Transport Systems, Berlin Germany, Oct. 1997.

[15] Decker, P., A. Salomäki, and S. Turunen, "GSM and UMTS as a Platform for Transport
      Telematics Applications," *4th Word Congress on Intelligent Transport Systems*, Berlin,
      Germany, Oct. 1997.

[16] GSM based ITS Services-ERTICO Strategy for Implementation, V. 1.4, ERTICO,
      Rue de la Régence 61, B-1000 Brussels, Belgium, Nov. 1997.

[17] Vieweg, S., GATS, "Global Automotive Telematics Standard and Example of Implemen-
      tation," *4th Word Congress on Intelligent Transport Systems*, Berlin, Germany, Oct. 1997.

[18] Lang, K. E., "Wireless Application Protocol," *4th Word Congress on Intelligent Transport
      Systems*, Berlin, Germany, Oct. 1997.

[19] "Online Motoring," Translated reprint from ADAC Motorwelt 04/97, *4th Word Congress
      on Intelligent Transport Systems*, Berlin, Germany, Oct. 1997.

[20] Scomazzon, P. and R. Andersson, "A high bit-rate data broadcasting system using the terrestrial FM radio network / SWIFT Eureka 1197 project," *EBU Technical Review*, No. 264, Summer 1995, pp. 4–12

[21] Kuroda, T., M. Takada, and O. Yamada, "Development of an FM multiplex broadcasting system having a large transmission capacity," *IBC*, Brighton, 1990, pp. 241–245.

[22] Isobe, T., et al., "Traffic Information Services using FM multiplex broadcasting," *IVHS Workshop on Communication with Vehicles*, July 1992.

[23] Kojima, H., T. Yamamoto, and Y. Kumagai, "VICS Information Broadcasting Service / FM Multiplex Broadcasting," 4th *Word Congress on Intelligent Transport Systems*, Berlin, Germany, Oct. 1997.

[24] EBU Statement D81-1996: RDS-TMC (Traffic Message Channel), *EBU Technical Review*, No. 270, Winter 199), p. 45.

[25] Parnall, S. J., "The longer term future of TMC," *Proceedings of the TMC-Now! Conference*, European Broadcasting Union, CP 67, CH-1218 Grand Saconnex (GE), Switzerland, March 1997.

[26] Parnall, S. J., "TPEG" - A strawman proposal for a stream format protocol for transport information, BBC Research and Development Department, R&D Technical Note 0747 (97)P, March 11, 1997.

[27] ADAC-The Active Partner of Road Transport Telematics, 4th *Word Congress on Intelligent Transport Systems*, Berlin, Germany, Oct. 1997. Also published in German in *ADAC Motorwelt*, Dec. 1997, p. 46.

8

Basic and Enhanced Radio Paging

8.1 Introduction

This chapter introduces the concept of FM Radio Paging. This technique offers the fastest way to provide a full paging communication coverage for a metropolitan area or even complete region or country. FM paging is growing quickly because it is cost-effective and it mostly uses an infrastructure already built for other communication purposes. FM Radio Paging puts the existing telephone system and FM broadcasting stations to more profitable use, providing a valuable service and generating revenue for the operator.

In addition to the paging system described in earlier versions of the RDS/RBDS standards referred to as "basic paging," the 1998 version of the new RDS/RBDS standards introduces the "enhanced paging" protocol while keeping full compatibility with the earlier one.

The aim of the enhanced paging protocol is to upgrade the battery lifetime of a pager, as well as to easily permit regional and international paging, and multioperator and multiservices operation.

8.2 What can be Achieved With Radio Paging?

The Radio Paging (RP) feature is intended to permit multiplexing of the paging signals (type 1A groups, 4A groups, 7A groups, and 13A groups—used only for the enhanced paging mode) into the (RDS) data stream containing any of the other RDS features. This technique permits the provision of Radio Paging using existing VHF/FM broadcasts as a transport facility, thereby avoiding the need for a dedicated network of transmitters and at the same time making use of the relatively large service areas of FM transmitters.

RDS Radio Paging has the great advantage that an existing VHF/FM broadcast network can be used to carry this service. This is especially true in developing countries, where it is difficult to implement a special radio transmitter network to only carry a paging service, as would be required for a POCSAG paging system. From an economics point of view, this is usually the most cost-effective solution for implementing a Radio Paging service, and for a broadcaster or transmission operator, there is also the opportunity to earn revenue from using RDS to provide such a service. The capacity of an RDS Radio Paging service is approximately 40,000 subscribers per broadcast network of transmitters. The capacity is strongly influenced by the mix of paging services that is provided to users. If the paging service carries only a bleep tone to alert the subscriber, the capacity will be higher; but if long alphanumeric messages are carried, then the capacity will, of course, be much lower since these require more paging data groups to carry the information.

Some of the countries already making extensive use of the RDS Radio Paging technique are France [1], Germany, Sweden, Bulgaria, Hungary, Slovenia, Croatia, the Czech and Slovak Republics, Ireland, Spain, Dubai, India, China, Indonesia, Russia, Ukraine, Tanzania, South Africa, Mexico, and the United States.

The full specification for the implementation of RDS Radio Paging is given in Annex M of the new RDS/RBDS specifications [2,3].

In comparison to former RDS/RBDS standards, significant enhancements were included, as follows:

1. A set of variable-length message types for numeric and function messages has been added to the RDS paging definition, enabling more flexible and efficient use of the transmission.

2. The new definition includes message types for both national and international paging. Previously, only fixed-length messages were defined for numeric paging.

3. A repetition flag has been added. This enables more intelligent processing of the received messages when messages are repeated to improve the reception reliability.

4. A paging call counter has been added. This is a running number for each individual message sent to a subscriber. This number permits the user to be alerted when a message could not be received.

5. The type 13A group concept was introduced for the enhanced paging mode.

6. A global addressing scheme was added for country, operator, and operator's service area management purposes, which permits several operators (15 operators per country; that is, operator codes (OPC), with 63 different service areas per country, which are paging area codes (PAC) to provide Radio Paging services for the same service area without any conflict among them). These codes are carried in type 1A groups, variants 0 and 2.

7. Roaming (international links) for international paging is also possible.

8. A paging network can now be defined without any connection to the PI code, as was previously the case. Only the country code component of the PI code is being used.

9. It is possible to design and manage a paging network completely unrelated to the radio programmes that the transmissions carry.

10. A new battery-saving mode has been defined that improves significantly (10 to 15 times) the power consumption of paging receivers (i.e., the pager is only fully powered during 0.3 seconds out of a two-minute cycle). This is achieved thanks to type 13A groups being placed at the beginning of the six-second interval, already defined in the basic paging mode. Battery life can be as high as up to 2,000 hours.

11. The address range of receivers was enhanced from one million to ten million subscriber address codes.

The enhancements described permit an easy upgrade of existing networks.

Radio Paging is intended to provide a broadcast paging service making use of the type 7A group for transporting the paging messages (see Figure 8.1), associated with the use of type 1A, 4A, and 13A groups (enhanced paging mode only). Subscribers to a paging service will require a special pocket paging receiver in which the subscriber address code is stored. Five types of call messages are possible:

1. A simple call (bleep tone) without additional message;

2. A 10- or 15-digit (international), or 18-digit numeric message;

3. An alphanumeric message of 80 characters, maximum;

4. Variable-length message;

5. Functions message.

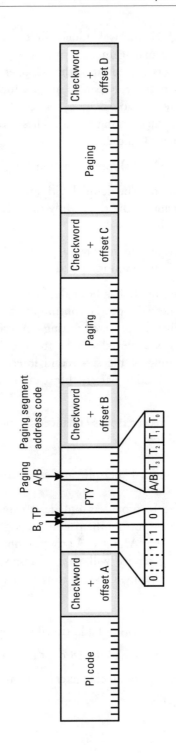

Figure 8.1 Type 7A groups convey the messages for Radio Paging, but other group types (1A, 4A, 13A) are also required for control and battery-saving functionality. (*Source:* EBU.)

In Figure 8.1, block 1 comprises the PI code found as the first block of every RDS group type. Blocks 3 and 4 are used to carry the paging message information. In block 2, the last five bits are used to control the paging information. Bit A/B is used as a flag that changes its value between different paging calls, thus indicating the start of a new or repeated call. Bits T_3–T_0 are used as a four-bit paging segment address code and to indicate the type of an additional message that follows.

8.3 RDS Paging Operational Infrastructure

The RDS paging network needs to bring together a paging service centre with the existing broadcast network to allow the paging data to reach the airwaves. Various configurations are possible, and Figure 8.2 shows many of the communication link options.

The public switched telephone network (PSTN) is used to relay request calls to the paging control terminal (PCT) and to send the required acknowledgements between the caller and the PCT. This requires the installation of a

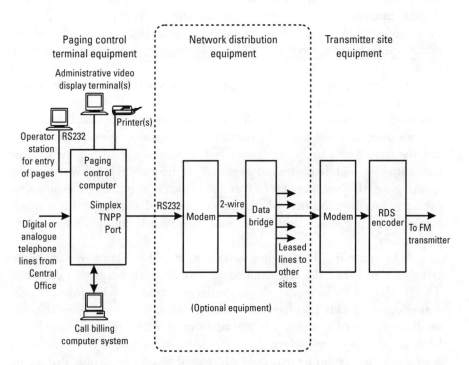

Figure 8.2 General description of an RDS paging system. (*Source:* EBU/RPA.)

special exchange designed for use in paging operations, performing the basic functions of organisation, control, and billing. These exchanges are commercially available from several manufacturers and can be located almost anywhere.

The paging message is processed with the addition of routing codes and other system management codes and then is packaged into an appropriate protocol to be carried via a suitable high-quality data transmission network to the input point or points of the FM broadcast network [6].

Mainly, two protocols are used. Originally the Telocator Network Paging Protocol (TNPP) was developed for wide area networks serving single-frequency systems, and it has also been used in RDS networks. The European Broadcasting Union (EBU) developed the Universal Encoder Communication Protocol (UECP), which in its latest version, 5.1, (see also Chapter 11), contains all the necessary commands for dynamically controlling RDS data to be broadcast. It also "transports" paging messages for the operation of both the basic paging mode and the enhanced paging protocol from the new RDS/RBDS standards [2–4].

It is important for the broadcaster or FM transmitter network operator to ensure system reliability, and this is best served by amalgamating all data into a single stream between studio centre and transmitter. These protocols will allow the use of satellite links, microwave data links, and data circuits by telephone cables equipped with modems for data transmission, or any combination of these.

The input to the FM network may be connected at the broadcast studio centre or at one, several, or all FM transmitting stations. If the input is to one transmitting station, all others will receive the transmitted signal, including the subcarrier from the original transmitter. It will also be possible to rebroadcast the transmitted multiplex (audio and data subcarrier) from the original transmitter by modulating it onto another VHF frequency. However, generally speaking, an RDS encoder may be required at each transmitter site to modulate the data onto the subcarrier when only the baseband audio signal is rebroadcasted. In each FM transmitting station, the required configuration may be very similar.

When operating RDS paging, it is necessary to provide an even more rugged data signal than the one required for mobile reception with a car radio. This is achieved by increasing the subcarrier deviation for RDS from the normal value of ±2 kHz to a level of at least ±4 kHz, the reason being the very small antenna used in a pager. In addition to this, each paging message has to be repeated two or even three times at a sufficiently spaced time interval of several minutes, assuming that most pagers will be used by people that are on the move.

8.4 Paging Receivers

It is undoubtedly a real challenge to design a small autoscanning paging receiver that is easily accepted by the user and that can fulfil all the requirements of the RDS/RBDS standards. RDS pager development has to be optimised by trading off size/weight, sensitivity (most critical for the provision of a reliable service), displays, alerts, and battery lifetime. Initially, the prices for RDS pagers were relatively high, but now they can even be as low as USD $100.00. The potential to compete with POCSAG pagers thus has grown dramatically. Important RDS pager manufacturers are now Info Telekom (see Figure 8.3) and Cirkisys. Nokia sold its paging business unit to Info Telecom and is no longer active in this sector.

The RDS system allows both numeric and alphanumeric paging options. Pagers can receive 10 to 18 digits and are equipped with a memory that permits storage of 10 messages or more. Alphanumeric pagers can display messages of up to 80 characters, and usually offer extensive memory capacity as well (see Figure 8.4).

Figure 8.3 The 1997 model RDS FM 101 of a new pager from Info Telecom in France, permitting reception of 10- or 18-digit messages. (*Source:* Info Telecom.)

Figure 8.4 The 1997 model RDS FM 110 of a new pager from Info Telecom in France, permitting reception of 80 alphanumeric character messages. (Source: Info Telecom.)

Each RDS Radio Paging transmission includes the country code and the group code of the pager, with which the pager synchronises itself. To preserve the battery life, the processor of the pager turns the pager on and off as required. The pager is only fully powered during the transmission of paging messages to its own interval.

8.5 Future Developments

All operators and manufacturers involved in RDS paging collaborate closely within the Radio Paging Association, which is part of the RDS Forum. These companies drive the technology and are thus the leaders in the worldwide market of FM Radio Paging. Apart from RDS paging, this market also uses some other technologies such as MBS [5], which is of special interest only to those who use FM broadcast transmitters that are operated outside the VHF/FM Band II (87.5–108.0 MHz); that is, mainly in countries of Central and Eastern Europe that still use the VHF Band I (66–74 MHz) for FM sound broadcasting and where RDS cannot be implemented.

Another proprietary system widely used in the United States and Canada is the system operated by Cue Network called MMBS, which consists of data multiplexed together with the RDS data stream using the offset word 0. This system can only be used in conjunction with the North American RBDS standard and is not included in the CENELEC RDS standard.

More recently, FM paging has also been implemented in a few cases on the DARC/SWIFT subcarrier system (see also Chapter 13), which offers a much higher usable data rate than RDS.

8.6 Conclusions

In a world where mobile communications embracing mobile telephone technology such as GSM are becoming a part of everyday life, one fact remains: there is invariably a cost versus coverage equation to be balanced for both operators and subscribers. Today, the RDS paging network can give the operator national and even international coverage when linked to other operators worldwide. RDS paging offers a very competitive communication service to potential customer groups who rely on keeping in contact within a wide area as they travel. Pocket pagers are easy to carry and have such a low power consumption that one single AA or two AAA battery cells can power them for as much as 1,000 hours or more. There are many other paging technologies around that seem to compete with RDS paging; however, with a recent sharp fall in prices for RDS pagers, the potential to compete with other paging technologies has grown significantly. This paging system is also ideal for the developing world, which really has little or no interest in investing in multimillion dollar infrastructures that RDS paging simply does not require.

References

[1]   Frossard, D., "RDS Paging in France," *EBU Review-Technical*, No. 245, Feb. 1991, pp. 22–28.

[2]   CENELEC, Specifications of the Radio Data System, EN 50067:1998. European Committee for Electrotechnical Standardisation, Brussels, Belgium April 1998.

[3]   EIA/NAB National Radio Systems Committee, United States RBDS Standard, Specification of the Radio Broadcast Data System (RBDS), Washington D.C, 1998.

[4]   EBU Universal Encoder Communications Protocol UECP-Version 5.1, European Broadcasting Union, 17A Ancienne Route, CH-1218 Geneva, Switzerland, Aug. 1997.

[5]   Swedish Telecommunication Administration (Televerket)-Paging Receiver for the Swedish Public Radio Paging System, 76-1650-ZE, 1976.

[6]   RPA, News Bulletins No. 1 (1995) and 2 (1996) of the Radio Paging Association.

9

Open Data Applications (ODA)

9.1 Introduction

This chapter describes the Open Data Applications (ODA) feature that was developed by the RDS Forum to meet a service provider requirement—to use RDS for data services that would be developed *after* publication and during the relatively long life of the RDS and the RBDS standards. A particular motivation was to facilitate the development of RDS-TMC (within the CEN standardisation processes) using ODA. But other uses have already emerged—some where a service provider wishes to make the use public and others where the service details are kept private, perhaps because it will be offered as a subscribed service.

9.2 The Concept and Availability of the ODA Feature

The ODA feature was developed to allow data applications not previously specified in EN 50067:1992 to be conveyed in an RDS data stream with minimal difficulty. The ODA feature, specified in EN 50067:1998, permits a number of predetermined groups types to be used, and these have to be indicated in any RDS transmission using the ODA application identification (AID) feature to allow decoders to monitor for them and then decode the relevant data. A wide variety of groups are available for use by ODA data services, and these may be selected by a transmission operator at his or her convenience. The service is identified in the associated type 3A groups in the *same* RDS data stream. Group types shown in Table 9.1 are available for ODA. Some can only

Table 9.1
ODA Group Availability, Signalled in Type 3A Groups

Group Type	Application Group Type Code	Availability for Open Data Applications
	00000	Special meaning: not carried in associated group
3B	00111	Available unconditionally
4B	01001	Available unconditionally
5A	01010	Available when not used for TDC
5B	01011	Available when not used for TDC
6A	01100	Available when not used for IH
6B	01101	Available when not used for IH
7A	01110	Available when not used for RP
7B	01111	Available unconditionally
8A	10000	Available when not used for TMC
8B	10001	Available unconditionally
9A	10010	Available when not used for EWS
9B	10011	Available unconditionally
10B	10101	Available unconditionally
11A	10110	Available unconditionally
11B	10111	Available unconditionally
12A	11000	Available unconditionally
12B	11001	Available unconditionally
13A	11010	Available when not used for RP
13B	11011	Available unconditionally
	11111	Special meaning: temporary data fault (Encoder status)

be used on the condition that other groups are not in use in the multiplexing of RDS groups of a particular transmission.

9.3   Indicating an ODA Transmission

Type 3A groups are used to indicate the inclusion of an ODA in any particular RDS data stream and inform a receiver/decoder about the ODA by providing both an AID and information as to which group the data service is carried in for the particular transmission being received. As a result, the receiver/decoder is able to invoke the correct software decoder for that application and monitor the chosen group type for data relevant to that ODA service.

The type 3A group shown in Figure 9.1 comprises three elements: the application group type code used by that application (using 5 bits), 16 message bits for the actual ODA, and the AID code (this uses 16 bits). Applications that actively utilise both type A and B groups are signalled using two type 3A groups. The type 3A group is usually transmitted at a relatively low rate, perhaps one group per minute, to provide information needed when a receiver/decoder is switched onto the relevant RDS data stream—either newly switched on or vectored from another channel. The data service provider has to decide how frequently the information contained in type 3A groups is needed, with the view of minimising the use of this group and avoiding taking up unnecessary groups. Generally, it is common practice to try to achieve about one type 3A group per minute for services that expect receiver/decoders to be used in nondedicated applications. However, closed user group applications with dedicated receiver/decoders may use a more infrequent use of type 3A groups.

The application group type code includes two special status conditions that may also be indicated by the following:

- The code 00000—"Not carried in associated group."
- The code 11111—"Temporary data fault (Encoder status)."

The former status is used to "prepare" a receiver/decoder by informing it that a particular software decoder should be invoked—perhaps some minutes before relevant data for a particular service will be transmitted so that the receiver/decoder can be fully ready for the data service when it becomes available. The latter code may be used by the RDS encoder to signal that incoming data to the encoder cannot be transmitted for some reason.

For each group type addressed by the application group type codes of a particular transmission, only one application may be identified as the current user of the channel. The AID code 0000 (Hex) may be used to indicate that the respective group type may be temporarily used for the normal features for which that group is already specified in the RDS standard. Application identification codes 0001 to FFFF (Hex) indicate applications registered in the ODA directory. These AIDs are allocated by the RDS registrations offices upon satisfactory receipt of a completed registration application form (as shown in

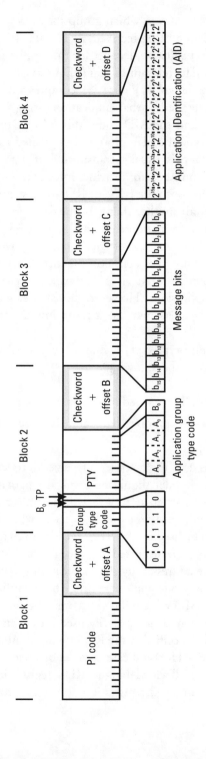

Figure 9.1 Bit map of type 3A group used for application group type code and AID. (*Source:* EBU.)

Figure 9.2). The registration office for Europe and the rest of the world, except North America, is at the EBU, Geneva, Switzerland. For North America, covering Canada, the United States, and Mexico, the registration office is situated at the NAB, Washington, DC.

The choice of group type for a particular transmission of a registered ODA will depend upon the other features used within the particular dynamic RDS multiplex. For example, if a transmission carries an in-house data stream in type 6A groups, then this group will not be available, so another group type should be chosen. But the *same* ODA may be conveyed in a *different* transmission using the type 6A group if it is not used for in-house data.

9.4 The Group Structure of Open Data Applications

Open Data Applications must use the data format shown in Figure 9.3 when using ODA type A groups and as shown in Figure 9.4 when using ODA type B groups.

Both type groups are available according to the data application needs. A large-capacity data service would use type A groups, which offer 37 bits per group, whereas a low-capacity service needing only 21 bits would possibly only need to use a type B group. For a very low data capacity service, there is a third option and that is to use no ODA groups, but just use the type 3A group, which has 16 message bits (see Figure 9.1).

Question	Information	Comment
Application name:		5 or 6 words, maximum
Application description:	Please use additional pages if desired	Give as much detail as possible
Open data mode: (see Table 9.2)		Choose one mode, only
ODA details, specifications and references:	Tick, if publication not permitted [ ] Please attach additional pages	Give all details, proprietary documents and references
Capacity requirement for both the ODA and AID groups:	Tick, if publication not permitted [ ] a) .... ODA groups/second b) .... type 3A groups/minute Please use additional pages if desired	Indicate: ODA groups/second and type 3A groups/minute. Describe any constraints

This Form will be published in full, except the last two answers, if specifically not permitted.

Figure 9.2 Part of the ODA registration form. (*Source:* EBU.)

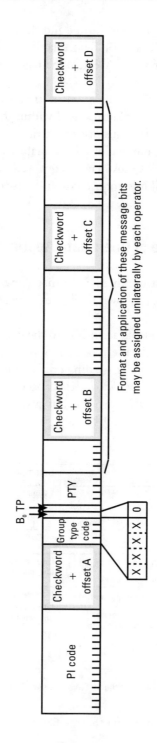

Figure 9.3 ODA type A groups, with 37 useable bits. (*Source:* EBU.)

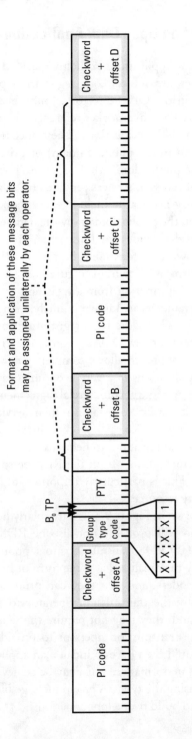

Figure 9.4 ODA type B groups, with 21 useable bits. (*Source:* EBU.)

9.5 Registration of an Open Data Application

There are no Open Data Applications explicitly specified in the RDS or RBDS standards. Instead, they are subject to a registration process. All registered applications are listed in the ODA directory, (published on the RDS Forum Web site (see Appendix K), which makes reference to appropriate companion standards and/or normative specifications. These specifications may either be public (i.e., specification in the public domain) or private (i.e., a proprietary specification not in the public domain). The terms "public" and "private" do not imply the degree of access to services provided by an application; for example, a public service may necessarily include encryption, so the details would not be made available in the public domain by the data service registrant.

The RDS standard provides for both situations, and a registrant may choose either situation without constraint. Part of the RDS Open Data Applications registration form is shown in Figure 9.2. Additionally, the form requires quite minimal information from a registrant to ensure that all ODAs are logged and basic parameters are known publicly. Generally, applications are confirmed quickly and easily by the RDS registrations office, which also gives advice on the requirements of new ODAs, in order to satisfy all the needs of the RDS specification. This approach allows completely new data services companies to develop new applications in a very economic way and to then be in a position to make the necessary commercial agreements with transmission operators anywhere they want to establish their services. To support this process, the RDS specification also carries a checklist that applicants must complete, together with their application, to help ensure that they have considered the best possible implementation—from the RDS perspective as well as from their service objective. This is particularly important to help preserve the very small RDS data capacity resource.

The checklist, shown in Figure 9.5, is particularly helpful in advising data service applicants (who will probably not also be RDS experts) to consider using other aspects of (all) RDS transmissions to support their service application. For example, they may realise that some form of frequency agility will be needed by receiver/decoders, and that they can probably use the AF feature already implemented by the transmission operator to achieve their requirement. On the other hand, they may not require the same network configuration as required by the transmission operator to provide service to an audio broadcaster, so they would then have to include an application specific regime to achieve the required performance. This may be achieved either in the ODA groups or perhaps by using the type 3A group message bits, since this information could be allowed to build over a longer period.

Question	Notes
Does the application behave correctly when not all RDS groups are received?	Necessary for mobile RDS applications
Does the application provide the means to identify the service provider?	
Does the application allow for future proofing, by upgrading?	
Does the application require sub-sets of associated applications?	Use of variant codes and/or other groups (eg clock-time)
Does the application include provision to reference other transmissions carrying the same service?	PI and AF
Does the application include an additional layer of error protection?	RDS already has considerable capability
Does the application include encryption?	
Does the application include data compression?	
Have you defined the capacity requirements for the application?	
Have you defined the capacity requirements for the AID under normal conditions?	
Is your application able to assume and lose the use of a group type?	
If so, have you defined the AID signaling when use of a channel is assumed?	
If so, have you defined the AID signaling when use of the channel ceases?	

Figure 9.5 Part of the ODA registration checklist. *(Source:* EBU.)

9.6 Guidelines for Using ODA

In order to further enhance the possibilities available to ODA designers, the RDS specification allows a full range of group type utilisations, as shown in Table 9.2. Normally, mode 1.1 or 1.2 will suffice, but it would be possible to mix group types if an application really required modes 2 or 3 to fully implement the required service. Clearly, these modes have an additional requirement by signalling through the use of *two* type 3A groups (using their application group type codes to fully describe the RDS datastream), that two ODA group types will be found in the service.

Table 9.2
ODA-Mode Options

Mode	Description
1.1	Type A groups, used alone
1.2	Type B groups, used alone
2	Type A groups and type B groups, used as alternatives
3	Type A groups and type B groups, used together

To ensure reliable data reception—for example, in the mobile reception conditions—it is recommended that data groups be repeated at least three times. The received blocks can then be compared bit-by-bit for acceptance, or if the message spans over several blocks or groups, one can use a two-byte cyclic redundancy to check in the last block for that whole message. If one byte in each group is reserved for message identification and as a counter, the groups could be mixed in any order and still be correctly decoded.

10

Differential GPS

10.1 Introduction

Using RDS for Differential Global Positioning Systems (DGPS) has become a reality in a large number of countries. However, although this issue has generated a relatively wide interest since the early 1990s—and implementation issues were widely discussed in both the RDS Forum and the U.S. National Radio Systems Committee—due to diverse commercial interests, no common standard has ever emerged. Ultimately, the new ODA feature was commonly identified as being most suitable to carry such a service.

This chapter describes the technological issues that need to be considered if one wants to implement DGPS within RDS. Such a decision would also have to take into account other alternative technologies that either achieve the same or even better results than those that one can now achieve with RDS. Internationally, continuous DGPS services that could, for example, serve European road users do not appear to be possible using RDS because of subscription and institutional challenges, which would require too much bandwidth to achieve satisfactory results.

In addition, one should also realise that whatever solution is adopted, it will probably only permit a relatively small market sector to be developed.

10.2 Positioning With GPS

The Global Positioning System is a worldwide satellite positioning/location system developed for military use by the U.S. Department of Defense (DoD).

It uses a network of 24 NAVSTAR satellites in nongeostationary orbits distributed over six planes; that is, with four satellites in each plane. The system was gradually built up and has become fully operational since 1995. It is currently U.S. government policy that GPS is also generally available for civil use and without any restriction to the international civilian user community. The technique is based on simultaneously measuring the distance from a GPS receiver up to at least four different GPS satellites. Due to various reasons, the distance measured to each satellite is not accurate. Nevertheless, GPS is able to deliver a standard positioning service that is limited to a ±100m accuracy level. However, the operator of the GPS system can influence the limits of accuracy through the introduction of a deliberate error source, known as the selective availability. The selective availability may be switched on or off for military reasons, and when switched off, the accuracy is increased to ±15m for civil use.

The satellites transmit coded information about time and their ephemeris (parameters that describe the orbits of the satellites). A GPS receiver (navigator) measures the propagation time of the signals transmitted and, with known signal propagation speed, the navigator converts the time to the distance between the navigator and each satellite (pseudo-range, PR), which then permits the calculation of its own latitude, longitude, altitude, course and speed (see Figure 10.1).

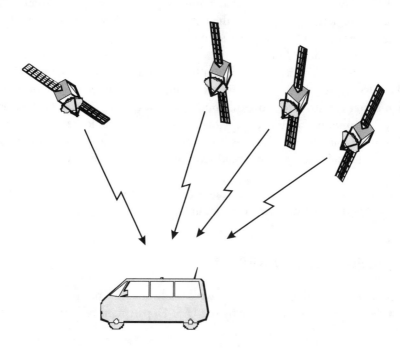

Figure 10.1 Principle of a satellite based positioning service. (*Source:* EBU.)

10.3 The Principle of Differential Correction

If there are two GPS receivers at different locations, each one registering identical signals, then the accuracy of the system can be augmented. For example, the baseline between the two receiving points can be computed with an accuracy of up to several millimetres when the data are combined for postprocessing. This technique is typically employed in surveying.

For use in real-time applications, a one-way data link is necessary between the two GPS receivers. Critical points are the data transmission rate and the distance between the two receivers. At low data rates (e.g., 100 bps), an accuracy of ±1–5m is possible up to a distance of 1000 km between the receivers. At higher data rates (e.g., 2,400 bps), an accuracy of several centimetres is possible up to a distance of 10–20 km between the receivers (see Figure 10.2).

The low data rates offered by RDS on FM or AMDS on LF and MF radio are thus suitable only for DGPS applications that are limited to the ±1–5m accuracy class. Higher accuracy applications require higher data rates and must therefore make use of DARC, DAB, or other telecommunication media.

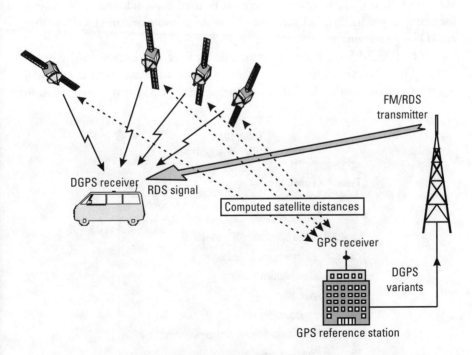

Figure 10.2 Principle of a DGPS positioning service using RDS. (*Source:* AZTEC.)

10.4   The RTCM DGPS Correction Format

10.4.1 Introduction

In 1983, the U.S. Institute of Navigation asked the Radio Technical Commission for Maritime Services (RTCM) to develop recommendations for transmitting differential corrections to users of GPS. A special committee, No. 104, was established, and it published version 2.0 of the RTCM Correction Format in 1990. A further developed version—version 2.1—was published in 1994 [1]; it contains additional formats that support real-time kinematic applications. RTCM version 2.1 provides 63 different message types, of which 26 were defined. Some of them are still tentative or reserved, while eight of the message types are fixed (see Table 10.1). Each type starts with a header, and the technical information is then allocated to a variable number of 30-bit words.

The RTCM SC-104 DGPS format is widely used by GPS manufacturers, and a significant effort was made in deriving that format. However, the issues dealt with by the RTCM committee are somewhat different than those that must be addressed when a data channel with relatively limited data transmission capacity such as RDS is used. The underlying principles are, of course, the same, and the RTCM format can therefore be used as a guideline to determine, for example, the RDS-ODA format to be used for implementing DGPS within an RDS data stream.

The RTCM format itself is unsuitable for RDS due to its excessive bandwidth. Therefore, the correction elements required will have to be compressed for transmission via the RDS data channel. The RTCM format then needs to

Table 10.1
Defined RTCM Message Types

Type	Function
1	Differential GPS corrections
2	Delta differential GPS corrections
3	Reference station parameters
5	Constellation health
6	Null frame
7	Beacon almanacs
9	Partial satellite set differential corrections
16	Special message

be reconstructed within a special service-provider-specific DGPS RDS FM subcarrier receiver before the correction data is delivered to the navigator. For example, to achieve a ±5m accuracy, 20 to 50 bps within RDS-ODA to carry the correction data will be sufficient. Remember, as explained in Chapter 9, one ODA group type A (e.g., type 11A group) can carry 37 bits for the application under consideration.

10.4.2 Required Data Elements

Differential GPS requires that certain types of information be available to the DGPS navigator [2]. In the following, some of the most important elements commonly used by DGPS service providers and based on RTCM formats are briefly explained.

10.4.2.1 Pseudo-Range Correction (PRC)

The pseudo-range correction is the most fundamental data item needed by the navigator. It describes the correction for a satellite that should be applied at a certain point of time (the time at which the pseudo-range measurements were made at the reference station). The PRCs are strictly only valid for the epoch when they are computed at the reference station. When the PRC is transmitted to a mobile user, time goes by, and the values grow older and deviate more and more from the actual value. To overcome this, a reference station makes a prediction about the rate of change of the PRC, which is called the range rate correction (RRC).

10.4.2.2 Range Rate Correction (RRC)

The RRC is used by the navigators to propagate the PRC forward to the point in time at which it can be applied by the navigator. Since the PRC was generated at a time in the past and a variable time was required for it to be transmitted and received by the navigators, the rate term provides the means for the correction to be made current at the navigator.

The composite PRC computed by the navigators from the PRC and RRC terms is in error to the extent that unmodelled effects cause the actual pseudo-range correction to differ from the computed pseudo-range correction. The most significant source of error is selective availability.

10.4.2.3 Age of Correction Data: Time of Correction with the Modified Z-Count

The correction PRC and correction rate RRC for a satellite are useless unless the times for which they were computed are also conveyed to the navigators.

In RTCM, the list of PRCs and RRCs for all satellites is therefore connected with the epoch of their creation, called "modified z-count," which counts the seconds of the current hour. A user DGPS receiver calculates the pseudo-ranges for the epoch in which it can use them. The predicted and broadcast values of RRC grow worse the older they are, such that accuracy degrades significantly with values older than 20–25 seconds.

10.4.2.4   Satellite Identification (SATID)

The correction information for each satellite is different, and there is no assurance that corrections are received by all navigators. Therefore, it is necessary to identify the satellite whose corrections are being transmitted.

10.4.2.5   Ephemeris Parameters

The ephemeris is a set of parameters that describe the orbits of the satellites. The ephemeris information for each satellite is changed periodically (approximately every one to two hours). The DGPS correction information is based on a particular ephemeris, available to the reference station when the correction is generated. It is necessary to advise the navigators of the ephemeris in use at the reference station when the corrections were computed, because the corrections may be properly used by the navigator only if the same ephemeris is employed.

The issue of data ephemeris (IODE) is the key that ensures that the user equipment calculations and the reference station corrections are based on the same set of orbital and clock parameters.

10.4.2.6   User Differential Range Error (UDRE)

It is important for the reference station to advise the navigators of the validity and approximate error in the differential corrections. This information permits the navigators to decide whether or not to use the correction from a particular satellite and how the satellite may be weighted in the position computation.

10.4.2.7   Reference Station Location Data: Station ID and XYZ Coordinates

In general, it is useful for the navigators to know the location of the reference station. This permits the navigators to use ionospheric and tropospheric models to estimate the amount of error in the pseudo-range measurements mode at both the reference station and the navigator.

This cannot be done at the reference station since the navigator locations are unknown to the reference station and there is no consistent model available for either troposphere or ionosphere that is accepted by all GPS manufacturers. For this reason, no ionospheric or tropospheric models are used at the reference station.

In the RDS case, it is assumed that the reference station is relatively close to the FM transmitter. The limited range of most FM stations (about 100 km) implies that the navigators will not be far from the transmitter. Hence, a user close enough to receive the FM station will generally be well within the range at which the differential corrections are accurate. These arguments indicate that there may not be a strong need for the navigators to know the location of the reference station.

However, the reference station altitude may be useful, because the atmospheric models that are commonly used require the altitude of both the reference station and the navigator. The tendency to site FM transmitters at relatively high locations to improve signal coverage tends to increase the need for reference station altitude.

10.4.2.8 Service Quality Parameter: SQ/Station Health

This DGPS format offers positioning service providers the possibility for services with different accuracies, depending on the frequency at which corrections are supplied. An indication in the messages is given to advise navigators of the expected accuracy of the service. The service quality parameter, Station Health provides this capability.

10.5 How RDS can be Used for Differential GPS

10.5.1 Design Considerations

For an average number of satellites, the type 1 RTCM message (the most important and most frequently transmitted message) is 500 to 700 bits long (e.g., 680 bits for nine satellites) and is split into independent parts. For a DGPS service within RDS, this data has to be rearranged so that the data concerning each individual satellite requires only 37 bits and can therefore be conveyed in one RDS group. Additional overhead information, which is necessary to reconstruct the exact RTCM message, is required only a few times per minute.

This rearrangement of the satellite data guarantees a stabilised evaluation of the data. As already noted, to transmit the original RTCM format takes a long time. By creating self-sufficient or autonomous groups, only the information relating to one satellite can be affected by transmission errors such as those caused by multipath propagation—the rest of the data remains undisturbed.

PRCs are not transmitted with their actual value, but are related to the last integer minute by means of the RRC. The mobile station has to reconstruct the current value of the PRC by using its own clock. Thus, the frequent transmission of the z-count can be avoided and the PRCs become nearly independent of any overhead information. The data for nine satellites need $9 \times 37 = 333$ bits, instead of the 680 bits needed by the original RTCM correction format.

When the RDS-ODA data format is used to transmit DGPS information, the AID and application group type code will be transmitted relatively infrequently in type 3A groups (i.e., once per minute). The DGPS correction data will then be mapped into the 37 available bits of the application group type (e.g., the type 11A group). Figure 10.3 [3] shows an example of how this is done in principle, that is taken from the German RASANT system [4]. To be able to determine the size for the RTCM data elements described above, it is necessary to decide how many bits will be required to map the information into the RDS-ODA application group. This decision implies that some kind of reduction for the RTCM data elements will be used that takes into account the available data transmission rate (limited in a practical implementation to only one or two RDS-ODA application groups per second) and the few corresponding bits thus available in these groups.

Each service provider has to decide this data format specifically for the service to be provided. Therefore, the DGPS RDS decoder that the end user wishes to use will always be specific to the particular service provider in spite of the fact that the GPS navigator uses standardised RTCM-formatted DGPS correction data. The DGPS RDS special receiver thus has to supply the correction data as input to the GPS navigator in the format of the RTCM standard.

Neither the RDS standard nor the RBDS standard contains an open standard for implementing DGPS. This is due to the fact that at the time when the RDS-ODA specifications were drawn up within the RDS Forum in 1995, DGPS using RDS as a data transport mechanism was already implemented in several countries—or at least at a very advanced state of system design. Several service providers were competing with similar but different proprietary data formats and advocating open and encrypted DGPS services. Because of this, no agreement was reached to include an open DGPS standard into the RDS/RBDS standards, both under revision at that time for the purpose of upgrading. All that could be achieved was that there was general agreement that RDS-ODA would provide a suitable means to achieve DGPS implementation within RDS.

10.5.2 Service Examples

10.5.2.1 RASANT in Germany

In May 1995, the Working Committee of the Surveying and Mapping Authorities of the Federal Republic of Germany decided to introduce a real-

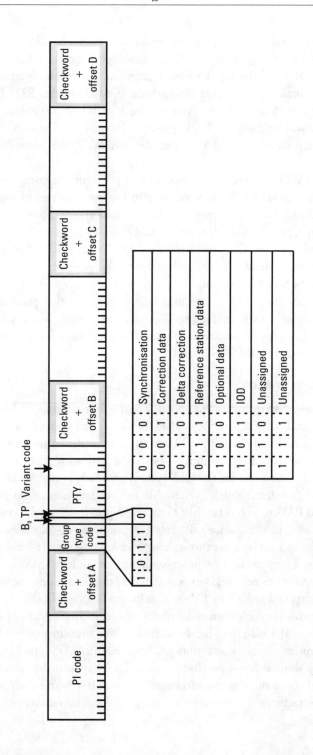

Figure 10.3 Example for using variant codes in block 2 to map 34 bits of DGPS data into group type 11A, blocks 2 (2 bits), 3 (16 bits), and 4 (16 bits). (*Source*: EBU.)

time DGPS service throughout the country. Radio Aided Satellite Navigation Technique (RASANT), which has been developed by the Westdeutscher Rundfunk (WDR) and the Landesvermessungsamt Nordrhein-Westfalen (the surveying and mapping agency of Northrhine Westfalia), uses RDS-ODA as the additional data channel. The service will be available from 1998 as an open service, and it will be financed by a one-time fee paid by the user, which is included in the price of the RASANT decoders available from several suppliers in Germany.

The RASANT technique can process and transmit the complete variety of RTCM messages via RDS. However, due to the low data rate of the present RDS service, RTCM message types 1, 2, 3, 5, 9, and 16 are modified and conveyed in an appropriate DGPS format, called DGPS variants.

Nearly all types of GPS receivers with DGPS capability can add these PRCs to their own pseudo-ranges and improve the positional accuracy from 100m down to between 1 and 3m. For the proven accuracy level of ±1–3m, an RDS capacity of one group per second is sufficient. In this case, the main DGPS variants, as shown in Figure 10.3, are supported. More capacity is required if additional variants are used (e.g., the alternative frequencies of the transmitters that carry a RASANT service).

10.5.2.2 EPOS in Sweden

Teracom, the Swedish transmission operator of the four national FM radio networks, introduced a DGPS service called EPOS at the end of 1994. EPOS is available as a basic service with a ±10m accuracy and as a premium service offering an impressive ±2m accuracy—both claimed with 95% confidence. The differential corrections for the GPS system are made available via the RDS channel all over Sweden through a renewable annual customer subscription and using modified RDS receivers (an EPOS receiver) purchased by the customers.

There are 12 EPOS service GPS reference stations, co-located with some of the Swedish land survey sites throughout Sweden to generate local differential corrections. These are first preprocessed and then all forwarded to a central concentrator and RDS network server situated at the operations centre (OPC) for the entire national radio and TV services located in Stockholm.

A multidrop network then redistributes the real-time stream of differential corrections to the matching local FM transmitter sites throughout Sweden. .At each transmitter site, the corrections are inserted into RDS type 11A groups. There is a very short delay—less than one second—between the calculation of the differential corrections at the various reference stations and their reception by the EPOS receiver. Integrity monitoring and backup functions for the

EPOS service are all incorporated in the system, and they are also supervised and operated by OPC.

The commercial service is much more than just the technical infrastructure described. Considerable effort was therefore invested in the design and implementation of a custom-made subscriber management system for the handling of customer subscriptions and other administrative functions. Gateways to Teracom's existing customer and invoicing routines were also established. The new data system even allows the customer—in addition to Teracom—to address and initiate his or her own EPOS receiver via the FM RDS network for the purchased service accuracy and subscription duration once the invoice has been paid. Distribution and sales of the EPOS service and the associated choice of receivers will be accomplished through cooperation with existing wholesale and retail outlets already selling GPS equipment.

10.6 Other Alternatives

10.6.1 Maritime Radio Beacons

The International Association of Lighthouse Authorities (IALA) has drawn up a frequency plan for maritime radio beacons to broadcast DGPS data. Implementations of this type already exist in several European countries where some 50 beacons are planned, with most of them operational since 1994/95 and broadcasting DGPS data with a bit rate of 200 bps. The coverage range of these transmitters is on the order of 200 km. The accuracy achieved is claimed to be between ±1m and ±5m.

10.6.2 AMDS

Since 1994, AMDS has been used in Germany on LF and MF, mainly on transmitters from DeutschlandRadio/DLF. AMDS uses a data rate of 200 bps, of which two-thirds is used to distribute DGPS data. The same service is also being tested in field trials conducted in Norway, the Netherlands, and Switzerland. Encoders and decoders are exclusively available from Bosch/Blaupunkt in Germany. Implementation expertise is available from Technische Universität Dresden, where the technology used was developed. The accuracy achieved is ±1m to ±5m. The coverage areas are very large. For example, the LF transmitter at Donebach uses 153 kHz and covers all of Germany, and Beromünster uses 207 kHz to cover all of Switzerland.

10.6.3 DARC

The relatively high net data rate of 9 kbps achieved with the LMSK-modulated DARC subcarrier on 76 kHz within an FM multiplex signal allows the provision of advanced GPS services, which may include a DGPS service with a high update rate and phase measurements that permit an accuracy of better than 10 cm. This was demonstrated during 1996/97 field trials in Austria involving the national broadcaster ORF. Similar tests, to be carried out in Norway, are also planned.[5].

10.7 GLONASS: the Alternative to GPS

Since 1996, the Russian Federation has been operating a system similar to GPS, called GLONASS. That system is also available for civilian use, but receivers are not yet as commonly available as for GPS. They use a different communication protocol. The accuracy without differential correction is around ±10m.

10.8 EGNOS: the European Component of GNSS

The Global Navigation Satellite System (GNSS) is an action plan launched by the European Commission. This plan has two phases [6, 7]:

1. GNSS-1 aims at implementing the European component, European Geostationary Navigation Overlay Service (EGNOS), which will use the U.S. GPS and Russian GLONASS signals with an augmentation (differential GNSS) of their integrity and accuracy by an additional ground- and space-based infrastructure scheduled to be in operation from the year 2000. This project is jointly pursued by the European Commission, the European Space Agency (ESA), and the European civil aviation control agency Eurocontrol.

2. GNSS-2 is envisaged as a new civil worldwide navigation system that will be internationally controlled and managed, and could be a successor of the existing military systems.

References

[1]   RTCM (1994)-Recommended Standards for differential NAVSTAR GPS service, Version 2.1, Radio Technical Commission for Maritime Services, Washington D.C, Jan. 3, 1994.

[2] Galyean, P., NRSC/RBDS DGPS Working Group—Status report of the coordinator, Torrance, CA, June 3, 1994, Working document of the RDS Forum, 1994.

[3] RDS Forum, Specification of the Radio Data System-Open DGPS, proposed additional specification, 1995.

[4] Raven, P., S. Sandmann, and G. Schoemackers, "RASANT Radio Aided Satellite Navigation Technique," *EBU Review-Technical*, No. 267, Spring 1996, pp. 27–31.

[5] Second SWIFT Forum, Lulea, Sweden, Published by Teracom Sweden, 1997.

[6] Global Navigation Satellite System, A brochure published in 1997 by the European GNSS Office at the European Commission (DG VII-A2) in Brussels, Belgium.

[7] Towards a Trans-European Positioning and Navigation Network, Including a European Strategy for Global Navigation Satellite Systems (GNSS)-Communication from the Commission to the Council and the European Parliament, COM (1998) 29/21.01.98

11

RDS Encoder Communication Protocols and the UECP

11.1 Introduction

This chapter explains the need for a communication protocol to be used between broadcaster (studios) or data service provider and RDS encoders. The EBU has developed a standard for this protocol that is commonly called the Universal Encoder Communication Protocol (UECP). It is especially recommended for use when RDS is implemented within a network of several transmitters and also when data needs to be assembled from various data service providers. The concept and the data format of this protocol are also described in some detail.

11.2 Why RDS Encoders Need a Communication Protocol

Transmission operators who want to implement an RDS service need to install RDS encoders. They are normally installed at transmitter sites adjacent to the stereo encoder, which generates a 19 kHz signal that the RDS encoder uses to synchronise its output RDS data stream.

Two "types" of RDS can be provided: Static RDS services can be provided by an RDS encoder simply providing RDS information, such as a fixed AF list, from internal memory; whereas dynamic RDS services, such as Radio-Text, require data input to an RDS encoder. Even static RDS services may have to be changed from time to time, depending upon the network configuration and the type of radio programme on air.

Dynamic RDS is needed for a number of reasons. RDS data related to the radio programme content (see Chapters 3, 4, and 6) requires a high degree of control from the on-air studio. In the case of a data service such as TMC, an operator-specific implementation under ODA, or Radio Paging (see Chapters 7–9), a high degree of control is required from the service provider to supply their data to the RDS encoder—either to a single broadcast transmitter or to a network of broadcast transmitters.

Simple but numerous different commands from the on-air studio or the service provider have to be sent, via a suitable data link, to the RDS encoder. For example, the on-air studio could change the status of the TP, TA, and MS flags to reflect the programme status. Such commands are often derived from "automated" functions to avoid operator intervention. A single-service broadcaster could possibly use a dedicated audio mixer channel for announcements, so that the MS flag is set to the "speech" condition every time the fader is opened, by using "back" contacts to send a command via the data link.

Once a dynamic RDS encoder has been chosen, then in most situations, a unidirectional data link is only possible for update data to be sent to the RDS encoder. This is due to the fact that for economic reasons, the transmitter operator had to adapt existing analogue, or even digital, audio distribution networks, which are essentially unidirectional, to add a data distribution function.

One possibility for implementing a unidirectional data link within an existing audio distribution network is using a subcarrier just above 15 kHz (nominally the highest audio frequency used in FM broadcasting), and using the fact that the audio channel is often designed for a 20 kHz bandwidth. A system of this kind is the AUDIODAT system, which was developed in Germany by ARD/SWF. This system has a bit rate of 200 bps. Communication codec equipment for this system is available from RE Technology and Rohde & Schwarz. This allows the audio and data signals to be combined at the studio, carried on the wideband audio circuit and extracted at the transmitter site, and for the audio to be connected to the stereo encoder and the data to be connected to the RDS encoder.

A number of proprietary update protocols were implemented on data links between source servers and the RDS encoders; several of them were encoder manufacturer specific. These protocols were used to send data messages from an RDS controller/management system (or simply an RDS encoder server) to the RDS encoders. Acknowledgements from the encoder were not essential; instead, it was arranged to send repeats of each message in order to ensure their receipt. In this way, a whole range of dynamic RDS features could be used by the broadcaster to enhance RDS performance for the listener. In some cases, an RDS encoder controller, running on a PC, has been used to send update messages to the RDS encoder. If a very full implementation is required, a dedicated RDS management computer system, frequently called an RDS server, has to be used to control on a scheduled basis the whole range of RDS features, including the radio programme-related data such as TP, TA, PTY, RT, and the full array of EON references.

11.3 Why the EBU and Encoder Manufacturers Developed the UECP

In the early 1990s, the EBU studied a requirement that the various existing and implemented RDS encoder communication protocols be harmonised. Such harmonisation would then enable broadcasters to purchase RDS system components (e.g., RDS encoders, RDS server computers, and software) from a variety of sources. This would permit significant economies in network operation, and it would offer the necessary high flexibility to implement in successive stages enhancements to already existing RDS implementations (specifically, within transmitter networks). RDS system component manufacturers would then also be able to integrate their products with those from other manufacturers, enabling more complex systems to be produced than those that would otherwise have been possible.

These proprietary update protocols had similar functional elements; however, they differed significantly in their environmental models. The structure, functionality, and addressing of their intended networks and the data structures within each RDS encoder are often quite different. Therefore, the Universal Encoder Communication Protocol (UECP) specification, now widely accepted [1], was based on *harmonised* environmental and encoder models.

The UECP is a layered communication protocol that is in line with the commonly used OSI reference model (ISO Recommendation 7498). The UECP in its present version—version 5.1 (August 1997)—encompasses all current RDS features, including the latest TMC specifications [2]. It is also designed to accommodate all new developments that will use the open data applications (ODA) concept (see Chapter 9).

The model and protocol also provide a template specification upon which new products may be based and, most specifically, it permits other existing encoder communication protocols to be enhanced. Thus, many existing devices can be adapted to meet the present functionality required. However, it is strongly recommended that all new RDS encoders should permit communications to also be operated under the UECP protocol (version 5.1 or later). The specification can now be freely obtained from the EBU by downloading it from the RDS Forum Web site (see Appendix K).

Organisations and manufacturers that have contributed within the EBU (and later within the RDS Forum) to the elaboration of the UECP specification, including AZTEC (see Figure 11.1), Auditem, BBC, Deutsche Telekom, Ericsson (formerly Teli), RE Technology, Rohde & Schwarz, TDF, Telefunken Sendertechnik, and Teleray. However, attention is drawn to the fact that at the end of 1997, only a few RDS encoder manufacturers had implemented UECP version 5.1. This version is specifically needed for TMC and ODA implementation.

11.4 The UECP Concept

11.4.1 Addressing Method

Communication to RDS encoders needs to be capable of many levels of addressing:

- To all encoders;
- To specific sets of encoders or to a particular device.

This may be accomplished by unique physical connections or, even better, by a suitable logical addressing method.

In defining an environmental model for the UECP, the following assumptions were made:

Figure 11.1 First RDS encoder implementation of UECP version 5.1: the AZTEC RADIOMEDIA FMX 440 Digiplexer (RDS and DARC). (*Source:* EBU.)

- The data stream will feed one or more transmitter sites. Each site will have a unique address, known as the *site address* (a number in the range 1–1,023). All encoders at a particular transmitter site share the same site address.

- An encoder will possess one or more site addresses. One of these must be unique to the particular physical site location. Additional site addresses are permitted for a particular area, region, or country.

To clarify this concept, an example is given. All encoders at the NEWTOWN site have the unique site address "123." Other encoders in the system are not permitted to use this address. Encoders at the NEWTOWN site also have the site address "267," which is allocated to all encoders in the LAKEVALLEY area. Messages arriving at the NEWTOWN site with either of these two site addresses will be accepted. Messages arriving at the LITTLEVILLAGE site (address "452"), also in the LAKEVALLEY area, will not be accepted if they carry the NEWTOWN site address, but will be accepted if they carry either the LITTLEVILLAGE or the LAKEVALLEY site address. This example describes the first level of the addressing system (see Figure 11.2).

A second level of addressing is now introduced, the *encoder address* (a number in a range 1-63). Several RDS encoders are installed at each transmitter site, serving a number of programme services. Backup equipment is sometimes provided, sometimes not. A single backup encoder may even be provided for several programme services. Whatever the situation may be, each encoder at the site needs to be individually addressable.

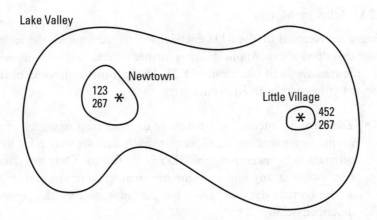

Figure 11.2 Fictitious example of site addressing with the UECP. (*Source:* EBU.)

An encoder will possess one or more encoder addresses. One must be unique to the encoder at that site. Additional encoder addresses may be assigned according to the encoder's usage or manufacture. However, the site and encoder addresses are not intended to specify a particular radio service. The specification of a particular radio service, a third level of addressing, is accomplished by using a *programme service number*. The site and encoder addresses should be thought of as being entirely physical, and are used only to address a certain "box" at a certain location. The functionality of the box is irrelevant in this context.

It is expected that many messages will be sent to all RDS encoders. Thus, the *global* number of zero (0) is defined for both the site and encoder addresses. Messages bearing the global site address are deemed to be acceptable at all sites in the system. Messages bearing the global encoder address are deemed to be acceptable at all RDS encoders at sites specified by the accompanying site address.

An RDS encoder will have two address lists, one of acceptable site addresses and the other of acceptable RDS encoder addresses. The site address list includes 0 (the global site address), the unique site address, and any additional site group addresses.

The RDS encoder address list includes 0 (the global encoder address), the unique encoder address, and any additional encoder group addresses.

A message is acceptable to a particular RDS encoder only if the site address is contained within its site address list and the RDS encoder address is contained within its RDS encoder address list.

11.4.2 RDS Encoder Conceptual Model

11.4.2.1 Software Model

Messages are accepted by the RDS encoder in accordance with the addressing method described above. Applicability is further determined by optional fields within the message itself (see Figure 11.3). This permits addressing of the following structures within an RDS encoder:

- *Data sets:* An encoder will have one or more data sets, each of which results in a particular RDS output. Each data set may refer to many programme services using the RDS EON feature. Only one data set is responsible at any one time for the encoder's output, and is known as the current data set. Data sets are addressed by the protocol as described below.

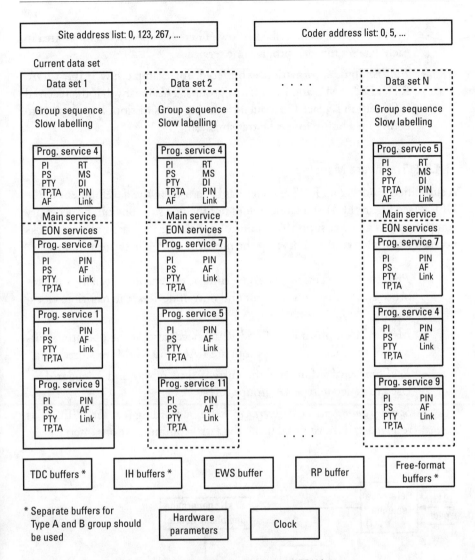

Figure 11.3 RDS UECP encoder software model. (*Source:* EBU.)

- *Programme services:* All programme services are identified by a unique *programme service number*, which is used to label data within RDS networks. In a network providing the EON feature, data for several programme services will be sent to an encoder, which may then identify that the data refers to one or more of the data sets and elements within the data sets used by that encoder. Programme services are addressed by the protocol as described below. There is a specific memory area in each data set for each programme service.

- *Buffers:* Some information is buffered, for example EWS, IH, ODA, RT, TDC, TMC and free format groups. This means that the received information is placed in a queue awaiting transmission. It is possible to configure a buffer for cyclic transmission.

11.4.2.2 Hardware Model

A simplified model of an RDS encoder has been used in the development of the UECP (see Figure 11.4). The model does not include such obvious or necessary components as a power supply or a control panel, but includes only the blocks necessary to understand and develop the protocol itself. These are as follows:

- *Processor:* The central processing unit of the encoder, usually a microprocessor, with access to input and output devices, the real-time clock, and memory.

- *Memory:* Comprises ROM and RAM necessary for the operating software of the encoder, and appropriate RAM and ROM for stored data.

- *Real time clock:* Maintains the current time of day and calendar date. Used to generate type 4A groups (CT).

- *Serial communication interface:* Data, according to the UECP, is received and transmitted using the serial communications interface.

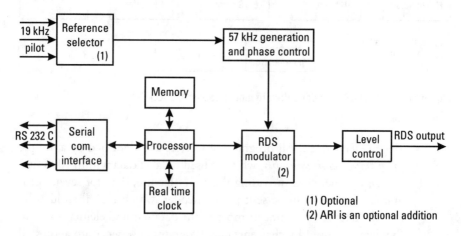

Figure 11.4 RDS UECP encoder hardware model. (*Source:* EBU.)

- *RDS modulator:* Produces the RDS biphase signal, and optionally ARI in accordance with the CENELEC EN 50067:1998.

- *57 kHz oscillator:* Frequency and phase locked to the third harmonic of the selected 19 kHz pilot-tone reference source.

- *Reference selector (optional):* Selects one source of a 19 kHz reference signal, out of a maximum of six, to lock to the internal 57 kHz oscillator.

- *Level and phase control:* The level and phase of the RDS signal (optionally ARI) may be adjusted by the processor under the appropriate commands.

11.4.3 UECP Transmission Modes

The UECP is designed to operate in various communication modes as follows:

- *Unidirectional mode:* This mode is used on one-way communication links. Data is transmitted to one, a group, or all RDS encoders. Answer back is not required.

- *Bidirectional mode, requested response:* This mode uses a two-way communication link to transmit data to one, a group, or all RDS encoders. It enables the server to request data, status, and error reports from RDS encoders.

- *Bidirectional mode, spontaneous response:* A two-way communication link enables a server to transmit data to RDS encoders and request data from RDS encoders. RDS encoders are also able to spontaneously generate status and error messages.

11.4.4 UECP Protocol Description

At the physical level, it is necessary for the UECP specification to ensure electrical and mechanical compatibility of equipment. Interfacing to the RDS encoder is accomplished with a serial interface based on the well-known standard EIA RS 232C (compatible with V24/V28). This is a full-duplex interface with hardware handshaking, able to operate with modems.

The data is transmitted byte-by-byte in asynchronous mode. Transmission speeds can be set at the standard values between 75 and 115,200 bps.

Update data comprise a stream of data frames. A frame comprises a series of bytes, delimited by two reserved bytes, FE hex and FF hex, which mark the beginning and end of the frame. Each frame contains a destination address, defining the set of RDS encoders to which the record is being sent. A sequence counter labels each separate record. The message itself is preceded by a byte, defining the message length, and after the message there may be a maximum of 255 bytes, followed by a CRC check word of two bytes.

The start and stop bytes, FE hex and FF hex, are uniquely defined and may not occur in any other fields of a frame. In order to prevent this, a frame is byte-stuffed prior to transmission. The technique of byte-stuffing allows a byte-oriented protocol such as this to preserve certain unique values for framing purposes, and yet allows conveyed messages to utilise the full byte range (00–FF hex). This is achieved by trapping reserved bytes in illegal fields and transforming them into legal byte pairs. Byte values FD hex, FE hex, and FF hex are trapped in the fields "Address" to "Cyclic Redundancy Check," inclusive, and transformed into a pair of bytes; that is, FD hex is transformed into FD00 hex, FE into FD01, and FF into FD02. Thus, the reserved bytes (FE and FF) will never occur within these fields in a transmitted record, and will only occur within the start and stop fields. When a message is received, the reverse technique is used to transform two-byte sequences (always starting with FD hex) into single bytes prior to the record being processed.

In cases where reserved byte values are present in the message, the transmitted message length will be increased. However, the length of a message is always defined in its unstuffed and shortest state.

Frames are built according to the structure shown in Table 11.1 and Figure 11.5, and then byte-stuffed prior to transmission. The check field consists of two bytes (prior to byte-stuffing), which represent the result of a 16-bit cyclic redundancy check (CRC) calculation. The divisor polynomial used to generate the CRC is the CCITT polynomial, $x^{16} + x^{12} + x^5 + 1$.

Table 11.1
Components of Each UECP Data Frame

Field Description	Descriptor	Field Length
Start	STA	1 byte
Address	ADD	2 bytes
Sequence counter	SQC	1 byte
Message field length	MFL	1 byte
[Message]	[MSG]	0...255 bytes
Cyclic redundancy check	CRC	2 bytes
Stop	STP	1 byte

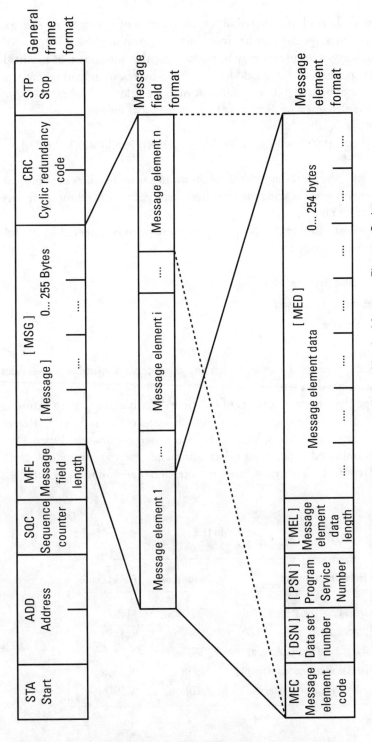

Figure 11.5 UECP Data protocol format consisting of a stream of bytes. (*Source:* EBU.)

[ ] Fields in brackets are optional. Inclusion is inherently defined by the Message Element Code

The message field, if nonzero in length, consists of one or more message elements. Each message element has the structure shown in Table 11.2.

Several message elements may be packed together into one message field, subject to a maximum message field length of 255 bytes. An individual message element must not be split between different message fields.

The complete message field may be represented as follows:

MEC,[DSN],[PSN],[MEL],[MED],[[MEC,[DSN],[PSN],[MEL], [MED]],...

Fields and whole message elements shown in square brackets are optional. Message elements may be concatenated freely, subject to a maximum message field length of 255 bytes.

The data set number (DSN) permits a message to be targeted to the following within an encoder:

- A specific data set;
- The current data set;
- All data sets.

The PSN permits a message element to operate a number of services within one or more data sets.

Message element data consists of one byte describing the length (number of bytes that follow as data) and the data, which are all coded as bytes. Different classes of message are defined, such as RDS message commands, Open Data Application commands, transparent data commands, paging commands, clock setting and control commands, and remote and configuration commands

Table 11.2
Format of Each Message Element

Field Description	Descriptor	Field Length
Message element code	MEC	1 byte
[Data set number]	[DSN]	0...1 byte
[Programme service number]	[PSN]	0...1 byte
[Message element data length]	[MEL]	0...1 byte
[Message element data]	[MED]	0...254 bytes

(RDS adjustment and control, ARI adjustment and control, control and setup commands, bidirectional commands, and specific message commands).

Remote and configuration commands make it possible to control the various functionality options of RDS encoders, or permit request messages to be given from the RDS encoders in the case of a bidirectional transmission mode.

RDS message commands (for more details, see Appendix L) make it possible to communicate all the RDS features that have to be processed by an RDS encoder.

The data is transmitted to the RDS encoder using the specified command structure and stored in memory according to the encoder software model. The RDS encoder must also be told about the types of groups to be transmitted and about the appropriate transmission rate for every transmitted group type. This is achieved with the group sequence command, which is treated by the encoder like a group enable command. When a specific group is encountered in the sequence, data relating to that type is transmitted, if available. If no data for a specific group type is received, then the group type is not generated and the next group type in the sequence is used instead. With this method, the desired repetition rate for every group type is implicitly defined in a very flexible way for the broadcaster or service operator.

References

[1]   EBU Universal Encoder Communications Protocol UECP, Version 5.1, European Broad-casting Union, 17A Ancienne Route, CH-1218 Geneva, Switzerland, Aug. 1997.

[2]   CEN ENV 12313-1, Traffic and Traveller Information (TTI)-TTI Messages via Traffic Message Coding, Part 1, Coding Protocol for Radio Data System-Traffic Message Channel (RDS-TMC) using ALERT-C, European Committee for Standardisation (CEN), 36 rue de Stassart, B-1050 Brussels.

12

RDS Demodulators and Decoders

12.1 Introduction

In this chapter, the functionality of a radio data decoder and the microcomputer that controls the PLL tuner is explained. Existing ICs from various manufacturers for the RDS decoder/microprocessor (MCU) are mentioned. Then, the implementation of RDS in the three main categories of consumer receivers—that is, car radios, home hi-fi, and portable radios—is discussed. Some known plug-in cards with an RDS radio for PCs are briefly presented, and the chapter closes with a section about professional software RDS data monitors and data analysers for PCs and laptops in particular.

12.2 General Principles

12.2.1 RDS Demodulator/Decoder Technique and Functionality

Figure 12.1 shows the radio data demodulator/decoder in relation to the other parts of a VHF/FM receiver equipped to use the Radio Data System. The decoder accepts the multiplex signal from the output of the FM discriminator as its input, and feeds its output to a microprocessor. This microprocessor drives a display, controls the tuning of the receiver, and also performs some of the later stages of decoding the radio data signal. Microprocessor systems are, of course, most commonly incorporated in many of the more sophisticated modern receivers, employing frequency synthesised tuning (PLL) to control these tuning functions and the presets chosen by the user. The RDS decoder

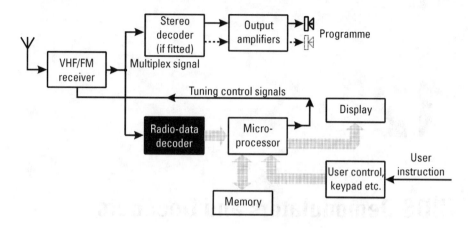

Figure 12.1 Typical block diagram of an RDS radio. (*Source:* EBU.)

can therefore be regarded as a logical extension. Such a control processor makes it possible to achieve better functionality and user friendliness, and eventually it will become an integrated part of the receiver.

Considering the functions of the radio data demodulator/decoder in a little more detail, six principal stages in the radio data demodulation/decoding process can be identified:

1. Demodulation of the 57 kHz suppressed-carrier amplitude modulated signal;

2. Decoding of the biphase symbols;

3. Recovery of the bit rate clock;

4. Recovery of group and block synchronisation;

5. Application of error detection and/or correction;

6. Decoding of address information and message codes.

All these stages can be implemented by a single microprocessor system. The techniques used to perform these functions are described in quite some detail in the relevant annexes of the RDS/RBDS specifications.

As an example of how this concept was implemented in a 1995 high-end car radio designed by one large European manufacturer, the following information may be of some interest. This example uses a powerful 8-bit single-chip microcomputer from Motorola's 68HC11 family (with PLL clock system, 32 KB of ROM, 1,024 bytes of RAM and 640 bytes of nonvolatile EEPROM, powerful timers, etc.) and a single-chip RDS demodulator with integrated

57 kHz bandpass from either Philips or SGS/Thomson to obtain and process the continuous synchronous data stream and the clock frequency of 1.1875 kHz (57 kHz/48). The relatively high power of the microprocessor is necessary because, apart from the RDS functions, it must control the PLL tuning system, the display of a car radio, the functional keys of the radio, the sophisticated man/machine interface, the digitally controlled audio processor, the cassette player, the remote controlled CD changer, the traffic announcement recording system, the clock, and so on. Nearly all these operations must be done in real time. Multitasking capability is therefore necessary for the different logical devices, because some devices must run simultaneously.

Every 842 μs (1000000/1187.5), the microprocessor will receive a new bit from the continuous RDS data stream delivered by the RDS demodulator chip, load it into a 26-bit software shift register, and process the RDS data in a 26-bit sequence. However, this is only true if the data stream is truly synchronous; that is, the position of each bit in the data stream can be clearly interpreted. Loss of only one bit means that the microprocessor is getting asynchronous data. In that case, a more time-consuming task needs to be done to obtain block and group synchronisation.

To acquire group and block synchronisation at the receiver (for example, when the receiver is first switched on, or tuning to a new station, or after a prolonged signal fade), the syndrome (see Appendix B) must be calculated for each received 26-bit sequence. That is, on every data clock pulse, the syndrome of the currently stored 26-bit sequence (with the most recently received data bit at one end and the bit received 26 clock pulses ago at the other) is calculated on every clock pulse. Then, with every new incoming bit (842 μs), a complete multiplication with the H-matrix (see Appendix B) has to be performed with all the 26 bits in the shift register. As a result, the 10-bit syndrome will be obtained. It must then be compared with the six known syndromes of the specified offsets A (block 1), B (block 2), C (block 3 for group type A), C' (block 3 for group type B), D (block 4), and E (occurs only with RBDS and detects MMBS data blocks in multiples of 4 in the RDS data stream). If a match occurs with a consecutive syndrome, 26 bits later (26 × 842 μs) after 22 msec, then the microprocessor is synchronised with the data stream. RDS synchronisation thus occurs when two consecutive RDS blocks in series have been detected.

If RDS reception quality is poor, this task requires a lot of processing power; but if the reception is good, the decoding task is, of course, much shorter. With mobile data reception, the RDS data stream can be full of errors, and a temporary 50% error rate is not uncommon. This is usually caused by the low RDS injection level used by the broadcaster (±1.2 kHz in some critical cases, while ±2.0 kHz is typical for the large majority of RDS implementations).

Another cause can be the simultaneous use of the ARI system (see also Appendix C, giving more detail about RDS data reception reliability). However, since the sequence of the syndromes is known, the decoding process can work as a "flywheel" mechanism, which is separately explained below. If an unknown syndrome is detected, the respective block is bad; but if block 2 is bad, the whole data group may be unusable unless error correction is used. However, the latter has to be used with caution because before it can be applied, the type of error has to be identified before it can attempt any corrections—and even then it can generate an error, which is less problematic for data used for display only (e.g., PS and RT).

It is very important to detect loss of synchronisation as soon as possible. One possibility is to check the syndrome continuously for acquisition of synchronisation. However, errors in the channel will make it difficult to continuously receive the expected syndromes, and therefore the decision must be based on the information from several blocks (e.g., up to 50 blocks). Another possibility is to check the number of errors in each block and base the decision on the number of errors in 50 blocks.

One possibility for detecting block synchronisation slips of one bit is to use the PI code, which does not usually change on any given transmission. If the known PI code is received correctly but is found to be shifted one bit to the right or to the left, then a one-bit clock slip is detected. The decoder can then immediately correct the clock slip.

12.2.2 Principles of the RDS Block Synchronisation System

The flywheel mechanism, used as a block synchronisation system in RDS, may perhaps be more easily understood with the aid of the simple mechanical analogy [1]. To develop that analogy, we first consider a conventional block synchronisation system, as illustrated in Figure 12.2(a). Here, the boundaries of blocks are marked by fixed block synchronisation words.

If we represent the received data stream as an endless belt, then these block synchronisation words may be represented as fixed markers on the belt, as shown. In the decoder, we must have a counter that is clocked at the bit-rate and that modulo the block-size. This counter may be represented as a slotted roller whose circumference is equal to the distance between the markers on the belt. It is easy to see that to synchronise the roller with the belt, all that is necessary is to allow the roller to slip until the slot meshes with one of the fixed markers. Once synchronised, the roller and the belt will remain in synchronism, even if some of the markers are damaged or missing. This illustrates the block-synchronisation flywheel system that is needed to enable the decoder to

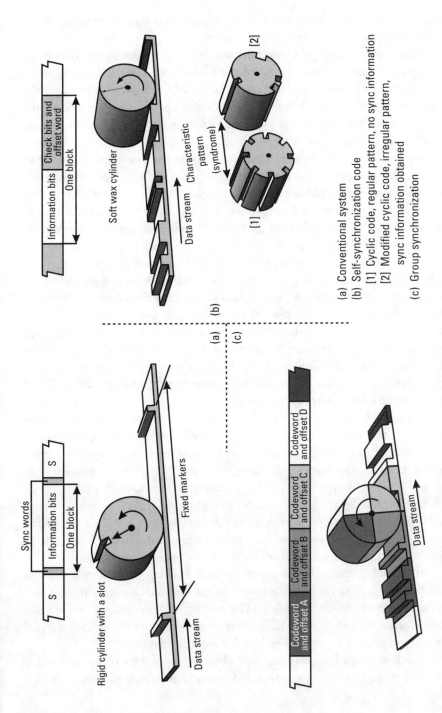

(a) Conventional system
(b) Self-synchronization code
[1] Cyclic code, regular pattern, no sync information
[2] Modified cyclic code, irregular pattern, sync information obtained
(c) Group synchronization

Figure 12.2 Mechanical analogies of the block synchronisation system. (*Source:* BBC/EBU.)

remain locked in synchronisation with the received data stream, especially when the latter is heavily corrupted with errors.

The disadvantages of a conventional block synchronisation system are also apparent from this analogy: Space must be allocated in the data stream for the markers, and these markers have to be clearly recognisable amongst the other data in the stream. Although these difficulties can be overcome (many data systems work this way), the particular requirements of radio data within an FM multiplex, favour a novel approach [2].

Using a similar mechanical analogy, the underlying principle of this novel block synchronisation system is illustrated in Figure 12.2(b). Here, the received code words are represented as a rack with *irregular* teeth, and the decoder in the receiver is represented as a soft-wax roller, which is capable of taking an impression of the rack as it is rolled along. Again, the circumference of the rotating roller is chosen to be equal to the distance between block (code word) boundaries on the rack. If the roller is rolled along the rack, starting at the beginning of the code word and ending at the end, then the resulting impression on the circumference of the cylinder will contain all the information about that code word. If the wax roller is subjected to further processing, analogous to the process in the error-protecting decoder, it is possible to see that the initial impression on the roller could be converted to one that depends not upon the transmitted message but only on the errors in the received code word. This pattern is analogous to the characteristic pattern or "syndrome" of an error-protecting decoder (see Appendix B).

If the movement of the roller is started and ended away from code word boundaries, the resulting first impression on the roller would comprise parts of two separate code words. It is a remarkable fundamental property of a conventional cyclic, or shortened cyclic code, that the characteristic pattern (syndrome) obtained by applying the error-checking process to such a mixture will, with high probability, be the same as for that obtained from one complete valid code word. In our analogy, we represent this case by the cyclically symmetrical pattern shown on the lower left-hand roller in Figure 12.2(b). Clearly, such a pattern would not provide reliable block synchronisation information.

If, however, the cyclic nature of the code is destroyed, which is done in RDS by adding binary offset words to each transmitted check word, an asymmetric pattern will result, as illustrated by analogy with the pattern on the roller in the bottom right of Figure 12.2(b). It is easy to see that such an irregular pattern can provide the necessary synchronisation information.

Block synchronisation is thus first obtained by trial and error, processing each 26-bit sequence until the expected characteristic pattern results. With a high probability, this will occur only when the processed 26-bit sequence comprises one complete code word.

Different offset words added to each check word will yield different characteristic patterns, and in RDS, four different offsets are used in each data group to identify the four blocks that form the group. This may be illustrated by developing the mechanical analogy, as shown in Figure 12.2(c). There, the circumference of the wax roller has been increased by a factor of four, compared with that of Figure 12.2(b). Then, after it has been rolled along the section of the rack that represents the four blocks comprising a group, four different characteristic patterns will result around the circumference of the roller, as illustrated by the different shades in that figure.

It should be noted that once one characteristic pattern has been found, the rest follows in cycle sequence. Thus, all the necessary group synchronisation information can be obtained within one block length, and this provides rapid data acquisition.

Here is a short summary of this novel group and block synchronisation system. The beginnings and ends of the data blocks may be recognised in the receiver/decoder by using the fact that the error-checking decoder will, with a high level of confidence, detect block synchronisation slip. The blocks within each group are identified by carefully chosen 10-bit binary words (offsets) that are added (modulo two) to the CRC check words such that offset word A is added to block 1 in each group, offset word B is added to block 2, and so on. For more details, see Appendix B.

12.2.3 Error Correction and/or Detection

The error-protecting code that is used in RDS is an optimal single-burst error-correcting code and is capable of correcting any single burst of errors that spans five bits or less. However, the use of the full error-correcting capability greatly increases the undetected error rate and thus also reduces the reliability of the block synchronisation system. In most experimental decoders used for RDS field trials, the following restrictions were observed when performing error correction:

1. When the decoder is searching for block and group synchronisation (e.g., on switch-on or on retuning the receiver to a different station), the error-correction capability of the code is not used at all. This allows the full error-detecting capability of the code to be used to detect synchronisation slip.

2. When the decoder has acquired group and block synchronisation, the error-correction system should be enabled, but should be restricted by attempting to correct bursts of errors spanning one or two bits. The encoder should attempt to detect (and then discard) blocks with longer bursts of errors.

12.3   RDS Integrated Circuits and Chip Sets

The major manufacturers of RDS components, having made them by the millions, are Philips Semiconductors and SGS-Thomson. The references for their current circuits are as follows:

- *Philips:* SAA6579T (demodulator and integrated 57 kHz bandpass filter). SAA6588 is an RDS preprocessor which, in addition to the demodulator and integrated 57 kHz bandpass filter, includes improvements resulting in more reliable data reception during weak signal input. Also, it performs RDS block detection, error detection/correction, synchronisation, flywheel, bit slip correction, and data processing control. CCR520S, CCR526S, and CCR610S are RDS-specific complete chip sets for high-end RDS/EON car radios that include the MCU with RDS-specific processing software. CCR991 and CCR922 are RDS-specific 8-bit MCUs with software for RDS/RBDS and MBS/MMBS (only CCR922) preprocessing.

- *SGS-Thomson:* TDA7330B or TDA7330BD, TDA7331B, or TDA7331BD, and TDA7479 (all these are single-chip demodulators with an integrated 57 kHz bandpass filter). TDA7332 or TDA73312D are 57 kHz bandpass filters. ST7285A5 is an 8-bit RDS-specific MCU (user ROM of 48 KB, data RAM of 3 KB) with an integrated RDS demodulator and a 57 kHz bandpass filter.

Other manufacturers in the years since 1987 were Bosch-Blaupunkt (in association with Texas Instruments), Sanyo (LC7070, LC72721, and LC72720N—single chips with filter and demodulator), Tokyo Cosmos, NEC (uPC2539GS—filter and demodulator), Siemens, and Rohm (BU1920F—filter and demodulator).

Prices have now become very competitive, and are generally at levels where the RDS functionality no longer adds significantly to the receiver price, except at the low end of the market.

12.4 Consumer Receivers

12.4.1 Car Radios

There are about 50 manufacturers in the European market with more than 400 different models, from high-end models down to the midrange and even down to low-end models, but the U.S. market still lags behind. We have to distinguish between first- and second-generation RDS; that is, in the first case only PI, PS name, AF, TP and TA flag features are supported, while in the second case, there is also EON (TA, AF, and PTY), PTY, CT, MS and even RT).

In the early stages of certain new products, especially from less well-known companies, some of the basic RDS features could have been left out and some malfunctioning of the RDS option was not uncommon. At a later stage of these products, improvements have been achieved, especially if the BBC's services for testing receivers were used.

Those manufacturers that have been in this market now for many years already have achieved, in the majority of cases, very fine products at very competitive prices, but to this category belong only, perhaps, a quarter of the 50 manufacturers. Fortunately, they have also made the largest number of the existing RDS radios sold.

In the United States, Delco, Denon, and Pioneer are among those companies that lead the RDS car radio market. They collaborate closely with the American carmakers to provide OEM products and in 1997, Ford, General Motors, and Toyota have all announced that they will start to offer American cars equipped with RDS radios (see Figure 12.3).

European carmakers have also started to offer their cars with RDS radios to the U.S. market.

12.4.2 Home Hi-Fi

There are now hundreds of models from about 25 different manufacturers on the European market. The U.S. market still lags behind. Prices range from very high to reasonably low. All support PS, some support AF, and several of them support RT, PTY, and CT. In this kind of product, the price of the RDS component is not significant anymore; therefore, we expect that in the future most of these models will be equipped for RDS.

12.4.3 Portable Radios

There is still some reluctance toward this kind of product on the side of the manufacturers, while broadcasters believe that, especially in this area, RDS

Figure 12.3 GM/Cadillac will be one of the first U.S. carmakers to offer RDS in 1998. (*Source:* GM/Delco.)

could significantly simplify the functionality and user friendliness of the radio for nontechnical people (imagine, as an extreme consumer case, your grand-mother), especially with the PI, PS and AF features. Of course, PTY and RT could also be very attractive options if they were more widely implemented by broadcasters, a situation that is likely to occur with the start of DAB implemen-tations in various parts of Europe. Some of the radio programmes to be found in the DAB ensemble will just be a duplication of the already existing FM programmes, but then it will be necessary to prop these programmes up with programme-associated data services such as PTY and RadioText.

Manufacturers that have portable RDS radio models on the European and U.S. market are Sangean (also sold as Uher or SuperTech), Siemens, Sony (notably the FM radio Walkman SRF-M48RDS, see Figure 12.4), and Grundig (see Figure 12.5). Prices are in the midrange with a trend toward less expensive models, with the cheapest being around USD $60.

12.5 Radios on Plug-In Cards for Personal Computers

12.5.1 The Philips SMART Radio

In 1996 in the United States, Philips Semiconductors showed a new chip set (TEA5757H, PCF8574T, TDA1308T, SAA6579T, CCR921, TEA6320T,

Figure 12.4 The Sony RDS Walkman radio SRF-M48. (*Source:* EBU.)

Figure 12.5 Grundig made the world's first portable RDS radio: the 1994 model YB 500. (*Source:* Grundig.)

TEA6360T, and TDA1517P) permitting the design and manufacture of add-on boards for IBM-compatible personal computers (see Figure 12.6). This is also available also as the OM 5604 multimedia FM receiver module, which is fully shielded to overcome the extreme electrical noise found in a PC environment.

Figure 12.6   Philips PC radio: software option 1. (*Source:* Philips.)

Philips has designed Windows reference software with Pascal source code to simplify the design task for the final product (see Figure 12.7) [3].

12.5.2 ADS Radio Rock-It RDS

In 1995, the Californian multimedia company Advanced Digital Systems (ADS) developed an FM stereo receiver/RBDS decoder PC plug-in card called

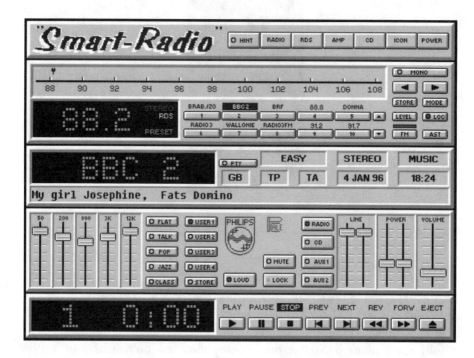

Figure 12.7   Philips PC radio: software option 2. (*Source:* Philips.)

"Radio Rock-it RDS." This unit has more than one application, being capable of displaying RBDS data or recording the broadcast programme to hard disk for later listening, and even includes a five-band graphic equaliser for enhancing broadcast or other input material. It is suggested that this system could be used to record the latest weather reports, allowing a weather-on-demand service [4].

12.5.3 The GEWI Radio G211 and TMC Office Decoder

The GEWI radio G211 is an RDS radio built into the small type 1 PCMCIA card. It comes with Windows software that installs a Window PC radio and an RDS decoder for data monitoring (see Figures 12.8 and 12.9).

The TMC office decoder is an additional piece of Windows software that enables TMC message decoding in conformity with the ALERT-C standard (see Figure 12.10).

12.6 RDS Data Monitors and Analysers for PCS

12.6.1 General Remarks

In this section, we present all the known professional RDS monitors and data analysers. Several of these products have been in existence for many years, and they have been upgraded several times to respond to many user requirements. They are all superb products on their own. However, our experience is that, for professional reasons, it is quite an advantage to have more than one of these software packages readily available and installed on the PC. The functions normally offered by each of these products are indeed numerous and, in terms of user-friendliness for any of the particular applications, there is frequently one or the other product preferred, which by no means decreases the value of the then nonpreferred product. There is, rather, a more subjective issue that the designer of such a software product could easily match for the individual case. Prices for

Figure 12.8 GEWI's PC card RDS radio with TMC software. (*Source:* GEWI.)

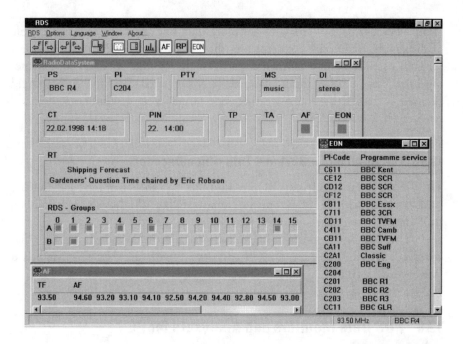

Figure 12.9 A screen image from the GEWI RDS decoder software showing BBC Radio 4 tuned and displaying the basic RDS features from that transmission, including PS, PI, PTY, MS, DI, CT, PIN, AF, EON, RT, and the groups used. Tiled over this display at the lower edge is part of the AF list, and on the right, part of the EON list showing PI and PS. (*Source:* EBU.)

these packages are nowadays very competitive and, as already indicated, the quality level achieved within these products is generally very high, so each one of them is in a way, a jewel in itself.

12.6.2 AUDITEM AUDEMAT Rx_MCRDS

This is Windows software (with optional language choice, English or French) for RDS data monitoring and analysis (see Figure 12.11). It comes with a kit that includes an RDS receiver unit with data output via a special interface cable that connects to the serial interface port of the PC.

12.6.3 AZTEC FM Explorer Version 3.0

This is a Windows software package for decoding and analysis of RDS data, radio frequency (RF), and audio measurements (see Figure 12.12). In addition, it needs an AZTEC RDS receiver connected to the PC. The FM Explorer

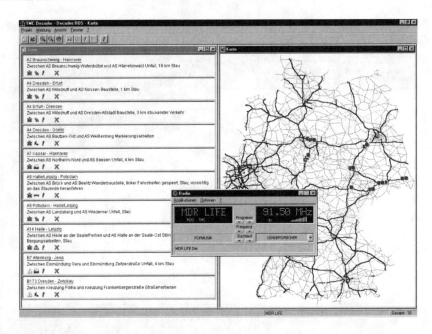

Figure 12.10 GEWI's PC card RDS radio with RDS decoder software and optional TMC software. (*Source:* GEWI.)

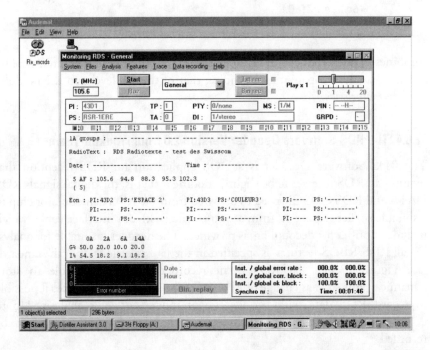

Figure 12.11 A typical screen from the AUDITEM AUDEMAT Rx_MCRDS. (*Source:* Auditem.)

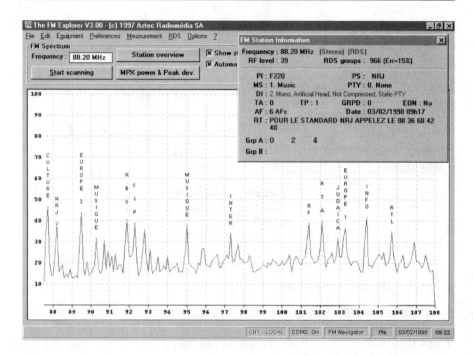

Figure 12.12 An FM band-scan made with AZTEC's analysis and decoding software. (*Source:* AZTEC.)

combines in one product a multitude of RDS functions with measurement and statistical analysis capabilities, including that of the RF level and the modulation characteristics.

12.6.4 The RDS Software Decoder Version 2.0 from Franken-Team

The RDS Software Decoder is a DOS application for RDS reception that requires an RDS receiver to be slightly modified; that is, the digital signals RDS DATA and RDS CLOCK need to be wired from the RDS demodulator chip to a suitable socket (e.g., a 3.5-mm stereo phone jack). An interface cable, which includes a buffer for decoupling, is provided. The decoder is suitable for analysing all RDS/RBDS features as specified in the latest versions of both standards (see Figure 12.13). An RDS event recorder makes it possible to store dynamic events such as change of PI or RT. The logic analyser makes it possible to record RDS data with various trigger modes for later analysis. Up to 2,100 groups of RDS data, including check bits (three minutes), can be recorded.

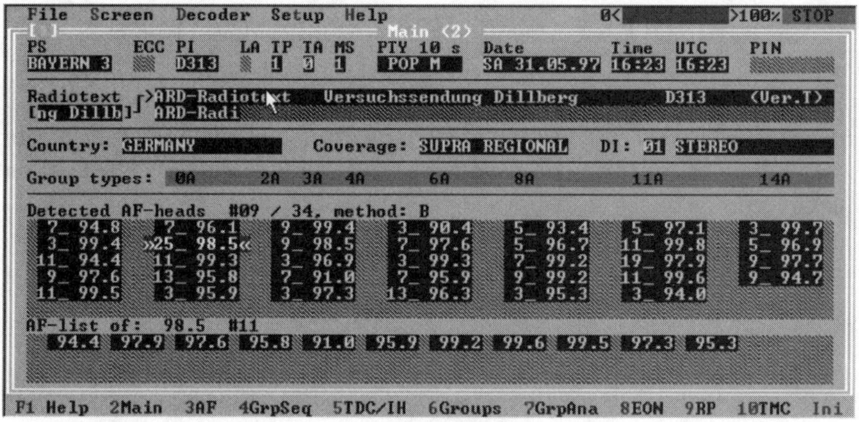

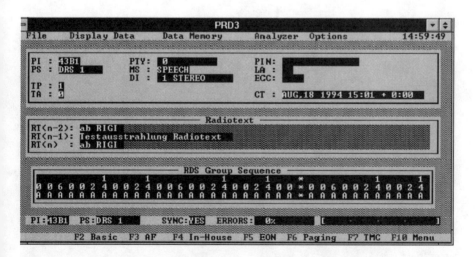

Figure 12.13 A typical screen from Franken-Team's RDS Software Decoder. (*Source:* Franken-Team.)

12.6.5 Schümperlin's PRD-3 and PRDLIB16.DLL

The PRD-3 is a DOS software program for decoding, monitoring, and analysing the RDS data flow (see Figure 12.14). The PRD-3-C comes with an interface cable that needs to be connected to the serial port of the PC and the RDS DATA OUT and GROUND pins of the RDS demodulator chip in an RDS receiver.

Figure 12.14 A typical screen from Schümperlin's PRD-3. (*Source:* Schümperlin.)

PRDLIB16.DLL is a dynamic link library for Windows (16-bit). It allows one to create specific RDS decoder applications in just a few minutes. Any programming language that supports DLLs can be used (e.g., Visual Basic, Delphi, C++, etc.). This DLL does all the low-level decoding of the RDS data stream (block and group synchronisation and error correction/detection). The user can thus concentrate on his or her own application.

References

[1] Ely, S. R., D. T. Wright, and C. C. Goodyear, "High-speed decoding technique for slip detection in data transmission systems using modified cyclic block codes," *Electronic Letters*, No. 3, Feb. 3, 1983, pp. 109–110.

[2] Ely, S. R., and D. Kopitz, "Design principles for VHF/FM radio receivers using the EBU radio-data system RDS," *EBU Review-Technical*, No. 204, April 1984.

[3] Philips Semiconductors, "Philips Smart Radio/Design Handbook," Application note, Nov. 1995.

[4] NAB, RDS Applications: Opportunities for Broadcasters, National Association of Broadcasters, Science and Technology Department, Washington D.C., ISBN 89324-229-2, 1995.

13

Outlook: RDS and Other Broadcast Data Systems for Radio

13.1 Introduction

This chapter reviews other radio data systems that are already used for data transmissions in the allocated frequency bands for terrestrial or satellite radio. The possibilities offered by these systems are compared with RDS wherever this is possible. A forecast is made about the market impact from the present point of view for all these systems. The following systems are being reviewed: AMDS and Digital Radio Mondial, the latter for AM broadcasting replacement by a new digital broadcast system; the high-speed data systems DARC/SWIFT, HSDS, and STIC for FM broadcasting; Digital Radio (Eureka 147-DAB), ADR, DSR, DVB, and WorldSpace.

13.2 LF, MF, and HF Broadcasting

13.2.1 AM Data System

Tests in the 1980s showed that low-rate data could be carried by LF and MF transmissions. For many years, the BBC 198 kHz transmitter has used phase modulation of the carrier to achieve 50 bps; however, tests on "Hochschule für Verkehrswesen Dresden" showed that it was possible to add a data stream of 200 bps using a phase modulation of $\pm 15°$ to an AM transmitter, with minimal degradation of the received audio quality. Since 1986, extensive field trials have been conducted to find an appropriate baseband coding to match the bit error conditions of AM channels in LF, MF, and HF broadcast bands.

The AM Data System, or AMDS, as it has come to be known, owes its origin to these trials. In 1994, the AMDS specification was expanded with the help of the EBU, but it concerns only the baseband coding for data broadcasting on LF, MF, and HF. Use was made of the experience gained by those broadcasters in Germany, France, the Netherlands, and the United Kingdom. They had conducted field trials with phase modulation as the modulation method, permitting a total bit rate of 200 bps. For example, in Germany, a national working party had already attempted to draft the specification, and the EBU made use of this to considerably widen its scope. In addition, in 1995, the ITU-R Working Party 10A prepared ITU-Recommendation BS.706-1 for AMDS. However, there are major problems with all the attempts made so far to get AMDS off the ground. One problem is that the modulation method used is incompatible with single-sideband broadcasting (SSB), which could in principle be used more and more for HF broadcasting. In addition, AMDS is incompatible with AM stereo, used in the United States. For these two reasons, the system has gained little support so far, and there are practically no suitable receivers apart from a few prototypes (mainly used for TMC, and DGPS field trials on LF and MF in Germany).

For HF broadcasting where changing ionospheric propagation conditions require broadcasters to use several HF bands during the course of a day, AMDS (as tested in these field trials) would definitely have been of some interest if SSB were not around the corner. In 1996, other alternatives were still being studied; for example, in the United States in the context of RBDS, and in Hong Kong by ID logic involving several large HF broadcasters from Europe and North America. Therefore, it is quite possible that a solution will soon present itself that overcomes the limitations that have so far hampered the success of an RDS for AM broadcasting.

The AMDS features described in the EBU's specification broadly follow the ideas of the RDS features, and many will be easily recognised, such as AF, DGPS, IH, RT, TDC, and TMC. Among the new features are additional tuning information (ATI), basic tuning and switching information (BTI), scheduling information (SI and SIS), and time information (UTC). SI and SIS pick up the idea of the EON feature of RDS and provide scheduling data for receivers to be stored in their memory. This feature is designed for HF applications, where programme breaks may occur in the order of every 15 minutes to 1 hour, resulting in many frequency and programme changes.

In 1996, there were still a number of test transmissions in LF, MF, and HF from European-based transmitters. One LF transmitter at 153 kHz for Deutschlandradio in Germany was providing TMC. Two MF transmitters in

Norway were providing DGPS to cover coastal waters, and one transmitter in Germany at 756 kHz was also providing DGPS. Receivable over a wide area were the HF transmissions from Deutsche Welle in Germany, at 6,140 kHz and 13,780 kHz in German and at several other frequencies carrying foreign languages.

13.2.2 New Developments Using Digital Modulation

Looking into the future of LF, MF, and HF broadcasting, it is more and more likely that analogue AM broadcasting will soon be replaced in these frequency bands with new radio systems using digital AM. These all will, of course, contain a radio data channel with features similar to those used in RDS.

At the time of this writing, the following steps in this development had already been completed [1]:

- *October 1994:* ITU Study Group 10 issues a report about digital AM;
- *November 1994:* First tests on MF in Berlin by Deutsche Telekom using SKYWAVE 2000 by Thomcast;
- *February 1995:* A European working party is formed to find a consolidated approach;
- *June 1996:* EUREKA accepts the new project EU 1559 Narrowband Digital Broadcasting (NADIB). Consortium members included Deutsche Welle, Radio France International, Telefunken, Thomcast, Sony, Deutsche Telekom, Teracom, and the Fraunhofer Institute;
- *December 1996:* Start of one-hour daily test transmissions on short waves from Jülich (Germany);
- *April 1997:* Establishment of the new collaborative project Digital Radio Mondial (DRM), led by Voice of America and aiming at the development of a digital short-wave broadcasting system that is resilient to fading, which characterises analogue shortwave broadcasting today.

13.3 FM Broadcasting

13.3.1 The Future of RDS

RDS will remain unchallenged in the market of FM broadcasting for many years to come, and today one can say with some confidence that no competitor to RDS is really on the horizon. The international scene for system standards clearly requires that other subcarrier systems used in FM broadcasting be fully compatible with RDS. This being a fact that cannot be easily changed, RDS will most probably remain in service as long as analogue FM broadcasting exists; that is, definitely for another 20 years.

RDS receivers have been marketed since 1987. The largest proportion have been car radios, and many of them are now also supplied as a standard-fit radio within a newly purchased car. The number of these receivers is steadily increasing, and there were more than 50 million in use by the end of 1997, as can be seen from Figure 13.1. However, many of these first-generation receivers only implemented the so-called five basic RDS features—PI, PS, AF, and TP/TA—for the simple reason that broadcasters using transmitter networks have first implemented these features for economic reasons.

In the second generation, these five basic RDS features were combined with EON. For some time, this situation has been changing, and we can already speak of a third generation. Broadcasters now give more attention to using the programme-related RDS features, such as RadioText and PTY, one of the reasons being that these will be used in DAB. Therefore, more and more RDS receivers now come onto the market that also implement these features. Originally, these were mainly home receivers, but now car radios will also implement RadioText. For safety reasons, however, there must be a switch with which the user can enable or disable RadioText—as a default, RadioText must be switched off in car radios.

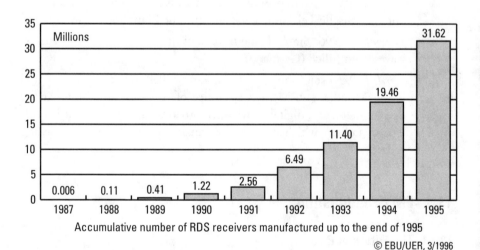

Accumulative number of RDS receivers manufactured up to the end of 1995

© EBU/UER, 3/1996

Figure 13.1 Development of the RDS receiver market. (*Source:* EBU.)

The large majority of consumer electronic manufacturers around the world support RDS technology. In Europe, prices for the RDS option are now only slightly higher than for a radio without RDS. There are already many manufacturers that equip almost all their FM radio products for the European market with an RDS module. Therefore, if the development continues with the same speed as in the past, we should soon reach the state where almost all new radios are equipped for RDS performance.

13.3.2 High-Speed Data Systems

13.3.2.1 General Considerations

The systems considered here use the frequency range above 61 kHz in the FM multiplex, and only those that are compatible with RDS will be mentioned. These systems achieve a significantly higher data rate than RDS.

Generally speaking, FM subcarrier systems appear, under certain conditions, to have some adverse effects on the reception of the audio carried on the parent FM transmission. This is particularly true in the presence of multipath propagation; that is in areas where the radio waves are bounced back by high-rise buildings or other obstacles, such as mountains. The significance and acceptability of these effects varies considerably, and depends on several factors: injection level of the subcarrier signal, its modulation system (where digital systems are less damaging), the particular type of receiver (fixed, portable, or car radio) and the way it is designed, the nature of the wanted radio programme (i.e., whether there is frequently low-level audio as in classical music or whether there is a continuously high-level sound) and the propagation environment (most critical are mountainous regions and cities with high rise buildings).

The addition of FM subcarrier systems is permitted only if they use all together no more than 10% deviation of the total FM carrier deviation permitted, which is ±75 kHz. If several subcarriers are used, it may be necessary to reduce the deviation for the main audio programme. Also, it is important to examine whether a subcarrier system complies with ITU-R Recommendations BS.412-7 and BS.450-2; that is, it has no adverse effect on the established protection ratios used for frequency planning.

13.3.2.2 DARC/SWIFT

Data Radio Channel (DARC) was developed by NHK and was, up to the end of 1997, mainly used in Japan (16 kbps) and partly also the United States (by a company called Digital DJ), occupying a bandwidth of 35 kHz centred on 76 kHz. The DARC injection level changes with the level of the audio programme modulation; that is, it is dynamic and depends precisely on the level of the stereo difference signal. The system characteristics are generally described in ITU-R Recommendation BS.1194-1 [2, 3].

In Europe, the DARC system is getting increasingly popular for the introduction of special open or subscription data services, which will be broadcasted to specific user groups. DARC was extensively tested and further developed by the Eureka 1197 Project "System for Wireless Infotainment and Forwarding Teledistribution" (SWIFT), which used the DARC modulation scheme (transmission layer 1 and 2) but a different baseband coding (transmission layer 3). This project started in 1994 and involved the telecommunications companies of France, Germany, and Norway and the broadcast transmitter operator, Teracom in Sweden. The project has produced a European ETSI standard that was adopted in 1997 [4]. First implementations are foreseen in Austria, France, Germany, the Netherlands, Norway, Sweden, and Switzerland, beginning in 1998.

To achieve a data service range as large as that of the audio programme, a higher data injection level is required (typically ±7.5 kHz) than the one used for RDS (typically ±2 kHz). Thus, when RDS and DARC are implemented, the maximum deviation of the audio programme has to be somewhat reduced to ensure that the total deviation remains within the ±75 kHz maximum permitted.

At the time of writing, the DARC system in Europe aimed at a niche market; that is, the provision of radio data channels to small portable data receivers (e.g., made by Casio, Sharp, and Sony) and PCs, with the receiver being implemented on a PC card (PCMCIA) using Japanese ICs from Sanyo or Oki. Use is also made of a conditional access technique with the key being electronically provided through a receiver identification code, permitting the user to be charged for the services provided. Services to be provided could be electronic newspapers, stock market and other financial data, an enhanced TMC, and the DGPS. Digital DJ (United States) started to launch similar services in the major U.S. metropolitan areas (see Figure 13.2).

Since 1996, the DARC technology has been promoted by the DARC Forum, which functions in a similar way to the RDS Forum, with some of the members participating in both.

The DARC Forum holds the view that the DARC system is a flexible and transparent data channel. It is an intermediate system that can be implemented in a short time in order to make the transition from RDS basic data services to advanced digital radio (DAB) data services possible by using the existing FM network infrastructure, which already provides excellent coverage everywhere. Thus, the DARC system can be easily implemented, providing a good coverage

Figure 13.2 The Digital DJ low-cost DARC receiver from Sharp is designed to deliver a new kind of subscriber radio data service in the United States, Europe, and Japan. (*Source:* EBU.)

area and a high availability for mobile reception at a fairly low cost. DARC could also be considered to provide a good marketing test opportunity for future radio data broadcast services [5].

13.3.2.3 Other Subcarrier Systems

Another system is the Seiko Telecommunication System (United States), HSDS, used on some transmitters in America and Europe. HSDS achieves 19 kbps, occupying the 57–76 kHz range in the FM multiplex with the subcarrier on 66.5 kHz.

A third system, so far of interest only in the United States, is called MITRE Subcarrier Traffic Information Channel (STIC) centred on 72.2 kHz (with a net user data rate of 7.6 kbps) from the U.S. company, Radio Dynamics.

In the United States since 1994, the NRSC High Speed Subcarrier Subcommittee has sought to evaluate the three systems with the view of achieving agreement on a voluntary U.S. standard that would recommend either one of these systems or a combination that incorporated all their recognised optimal features after an agreed upon field evaluation under identical test conditions for the three systems (DARC, HSDS and STIC). In March 1998, this objective was declared to have ended up in a deadlock due to the fact that the interested parties (mainly the system proponents) could not agree upon a single solution. The results of the field tests, which revealed that all the three systems performed relatively well, were submitted to ITU-R Study Group 10. This group decided to incorporate the results into ITU-R Recommendation BS.1194-1, which now recommends all three systems.

13.4 Digital Radio DAB

13.4.1 Origins and Possible Evolution

The future is most probably set for DAB, which already is increasingly used on VHF and the L-band (1.5 GHz). DAB is a European standard of ETSI (first published in 1994) [6] and a world standard recommended by the International Telecommunications Union (ITU). The system has been developed over a period of some ten years (beginning in 1986) by the Eureka 147 Consortium, with the EBU coordinating the writing of the specification. DAB is already widely implemented in large parts of Europe and Canada. Several other countries in the world also experiment with this system. In early 1998, about 150 million people were potentially able to receive DAB. However, receivers were still expensive and therefore largely unavailable except for field trials. However, expectations in the WorldDAB Forum are high, and indicate that this is going to change in late 1998 [7, 8]. Most recently, some important technological breakthroughs were achieved. A single chip receiver was developed jointly by Hitachi and Roke Manor, and a DABman (DAB version of a Walkman) was developed by Bosch. Clarion has developed a fully integrated one-box car receiver. In addition, a small British software company, RadioScape, has developed a full software DAB receiver that uses a standard Intel Pentium chip PC for decoding, rather than dedicated integrated circuits specifically designed for DAB. This opens the door to a wide range of multimedia applications that could be implemented via software, downloadable to the receiver; for example, a laptop computer being able to process the complete 2.3 Mbps DAB multiplex in real time.

DAB has numerous advantages over VHF/FM broadcasting. It offers significantly better reception quality for both audio and data services; that is, it overcomes the adverse effects of multipath propagation, so badly experienced in VHF/FM broadcasting. DAB offers a high degree of ruggedness. In addition, DAB provides for a very high spectrum economy due to the single frequency network technique and, further, a high flexibility for assembling, even dynamically, a number of different audio and data services to the service multiplex, called the "ensemble." One ensemble would typically provide capacity for six stereophonic audio programmes (each with a bit rate of 192 kbps), each having its own programme-associated data channel PAD (32 kbps) and additional capacity for other radio data in both the fast information channel (FIC) and the auxiliary information channel (AIC) [9–11].

The choice for multiplexing the ensemble is so wide that in the extreme, all audio could be replaced by data, permitting one video channel with sound.

This is a key feature of DAB that permits broadcasters to "flex" the multiplex; that is, the bit rate allocated to individual services in the multiplex can be changed on a dynamic basis to meet the needs of the broadcaster and the public. This powerful DAB feature thus opens the door to multimedia radio with several new data services that are still being defined. For example, for DAB, a media object transfer (MOT) protocol has been proposed for the transmission of multimedia services (see Figure 13.3). However, in the already existing definitions, much effort has been made to preserve compatibility with RDS features wherever possible.

Of course, many broadcasters will for many years need to distribute their services in both VHF/FM and DAB. Nevertheless, with DAB, existing RDS services can develop further and improve. For example, the Traffic Message Channel (RDS-TMC) will most probably be significantly enhanced with the new bearer-independent protocol TPEG, as already mentioned in Chapter 7, to carry urban and interurban traffic and travel information, including public transport and tourist information.

13.4.2 Comparison of RDS and DAB Data Features

The data services in DAB are being called service information (SI), and those already defined in the DAB standard are being summarised in Table 13.1.

Figure 13.3 DAB makes it possible to implement multimedia services on DAB. During a 1997/98 field trial in Germany, SDR/SWF used the MOT protocol to broadcast pictures with CD covers of pop music titles on air. (*Source:* EBU.)

Table 13.1
DAB Service Information (SI) features as compared to RDS

DAB Feature	Position of Feature in DAB Multipex	Transmission Capacity	RDS Comparison
Service-related data			
Service directory (part of MCI)	FIC	High	PI, ECC
Service label	FIC	High	PS
Frequency information	FIC	Medium	AF
Conditional access	FIC MSC	Medium	—
Programme-related data			
Programme language	FIC	Low	Lang.
Programme TYpe	FIC	Medium	PTY
Programme number	FIC	Medium	PIN
Foreground / Background	Audio	Low	—
Programme associated data (PAD)			
Dynamic range control	PAD	Low	—
Dynamic label	PAD	Medium	RT
Music / Speech / Off	PAD	Low	MS
Copyright	PAD	Low	—
Interactive text transmission system (ITTS)	PAD	Medium	—
Non-audio features			
Time and date	FIC	Low	CT
In-House	PAD/FIC	Low / Medium	IH
Transmitter identification	FIC/AIC	Medium	—
Paging	FIC/AIC	High	RP
Traffic Message Channel (TMC)	FIC/AIC	Medium	TMC
Emergency Warning Systems (EWS)	FIC/AIC	Low	EWS
Ensemble-related data			
Ensemble label	FIC	Very low	—
Announcements	FIC	Low	TP, TA
Alarm	FIC	Very low	PTY = 31
Other ensembles and services			
Other ensembles and AM / FM	FIC/AIC	High	EON

Reference is also made to how they correspond to RDS features; however, usually they are enhanced. An example is the service label ("BBC Radio One"), which has sixteen characters at maximum in DAB, but has only eight at maximum in RDS ("BBC R1"). The service label in DAB is thus eight characters longer than the PS name used in RDS. It was designed to give greater freedom in describing a service. To maintain compatibility with RDS, the service label is accompanied by additional information that allows up to eight characters to be suppressed in a receiver with a limited display. All features using a text display in RDS and DAB use the same character tables as those detailed in Appendix I.

The frequency information feature provides a cross-reference to alternative sources of service. It is similar to the RDS AF feature, which provides a list of alternative frequencies and permits programme continuity when a mobile receiver moves across coverage area boundaries and requires retuning to obtain better reception. Although DAB is suited to operation in a network where *all* transmitters operate on a single frequency network (SFN), DAB multiplexes may also be available via satellite or cable, and these multiplexes may operate in different frequency ranges and DAB modes.

Data capacities for DAB data services are significantly higher than is possible with RDS, and in Table 13.1 they are indicated in only very broad terms because the requirements depend on a wide range of options. "Very low" means a few bits per second and "high" indicates 10 kbps or more.

13.4.3 RDS/DAB Interoperability

When DAB services duplicate those radio programmes available on VHF/FM, and when DAB signals are not yet widely available (especially during the years when new DAB services are being introduced), there will definitely be areas where some services will only be available on VHF/FM. Then, the DAB frequency information feature will have to provide for cross-referencing to FM (and in theory, also to AM), and it shall consequently support the full range of radio frequencies foreseen for DAB, AM, and FM.

In the years 1996–1997, the RDS Forum established a working group to examine how various radio data features of the RDS and DAB systems could best be implemented by both broadcasters and receivers to allow a user to benefit from both systems used in his or her receiver. This was done with the view of avoiding confusion for the technically uninformed listener. Neither system should be modified, the only objective being to better understand how the system features should be applied to achieve these objectives. The findings of the working group are, at the time of writing, being considered for inclusion in the RDS guidelines that the EBU and the RDS Forum intend to publish during 1998.

13.5 Digital Radio by Satellite

Satellite radio is currently available in Europe via the digital satellite radio (DSR) system and the SES/ADR (ASTRA Digital Radio) system. DSR was only used in Germany, and it was unsuccessful in capturing a European market because it was conceived in the early 1980s for fixed antenna reception only. In the DSR implementation phase, starting in 1985, DAB was already being developed, and contrary to DSR, it was conceived for mobile reception while permitting fixed reception as well. This made DSR unattractive and, finally, redundant.

Some years later, in 1994, SES introduced on its ASTRA satellites the ADR system. Because of its limitation of fixed reception only (a similar limitation to DSR), it is likely to be around for only a few more years. ADR uses an analogue television signal to carry in a piggyback manner a modest number of radio channels as well as TV. In those European countries where the ASTRA satellite system is popular, ADR is inexpensive as a transmission medium, and receivers are available at a marginal cost increase to the ASTRA satellite TV decoders, which are generally available at low cost. ADR can be operated to include an in-house data channel that makes use of the Universal Encoder Communication Protocol, as described in Chapter 11. This permits the feeding of a network of terrestrial radio transmitters, where each individual transmitter can be fed in a regional context with RDS data as required for the terrestrial operation.

Another system to come soon is DVB-S. Intended primarily for digital television, it can provide radio to smaller sized dishes than ADR, but these dishes still have to be fixed and have a clear line of sight to the satellite.

The year 1998 may be a landmark year for digital radio by satellite, because a new radio system, called WorldSpace, will be launched and another digital radio system, Mediastar, may come in three years' time.

WorldSpace aims at providing low-cost digital satellite radio to portions of Africa, the Middle East, Latin America, the Caribbean, and Asia. The transmission system will not use DAB, but a proprietary system instead, details of which are not yet in the public domain. Two of WorldSpace's satellites are expected to be launched in 1998, with service beginning in late 1998 or early 1999. Later, the system may be used for the United States as well. It has little significance for Europe, at least for now.

Mediastar, on the other hand, is intended to provide Europe with satellite radio using DAB, with operation planned to start in 2001 using the 1.5 GHz frequency range. Three low-orbit satellites are planned, to be positioned over the target coverage areas. However, what impact these systems will have on the future of radio is generally unknown.

References

[1] Schall, N., "Digital developments in Amplitude-Modulated broadcasting," *Proceedings of the FES/ITU/PRO AV AFRICA/URTNA Regional Symposium in sound and television broadcasting for Africa*, International Telecommunication Union, Geneva, Switzerland, 1997, pp. 459–463.

[2] Kuroda, T., et al., "Transmission scheme of high capacity FM multiplex Broadcasting System," *IEEE Trans. on Broadcasting*, Vol. 42, No. 3, 1996, pp. 245–250.

[3] ITU R Recommendation BS.1194-1, System for multiplexing FM sound broadcasts with a sub-carrier data channel having a relatively large transmission capacity for stationary and mobile reception, International Telecommunication Union, Geneva, Switzerland, March 1998.

[4] ETSI, ETS 300 751 Radio broadcasting systems, System for Wireless Infotainment Forwarding and Teledistribution (SWIFT), ETSI Secretariat, F-06921 Sophia Antipolis Cedex, France, 1997.

[5] Andersson, R., "The possibilities of using DARC/SWIFT for datacasting," *Proceedings of the Montreux International Radio Symposium*, 1998.

[6] ETSI, ETS 300 401 Radio broadcasting systems, Digital Audio Broadcasting (DAB) to mobile, portable and fixed receivers, ETSI Secretariat, F-06921 Sophia Antipolis Cedex, France, 1996.

[7] Van de Laar, H., et al., "Towards the next generation of DAB receivers," *EBU Review-Technical*, No. 272, Summer 1997.

[8] Müller-Römer, F., "DAB Progress Report 1997," *EBU Review-Technical*, No. 274, Winter 1997.

[9] Plenge, G., "DAB - A new sound broadcasting system - State of the development - Routes to its introduction," *EBU Review-Technical*, No. 246, April 1991.

[10] Schneeberger, G., "Datendienste mit DAB," Deutsche DAB Plattform e.V., Am Moosfeld 31, D-81829 München, Germany, 1996.

[11] Lauterbach, T., *Digital Audio Broadcasting: Grundlagen, Anwendungen und Einführung von DAB*, Franzis-Verlag Feldkichen, Germany, ISBN 3-7723-4842-4, 1996.

Appendix A
Modulation of the RDS Data Signal

A.1 Subcarrier Frequency

The RDS data signal is carried on a subcarrier which is added to the stereo multiplex signal (or monophonic signal as appropriate) at the input to the VHF/FM transmitter. Block diagrams of the data source equipment at the transmitter and a typical receiver arrangement are shown in Figures A.1 and A.2, respectively.

During stereo broadcasts, the subcarrier frequency will be locked to the third harmonic of the 19 kHz pilot-tone. Since the tolerance on the frequency of the 19 kHz pilot-tone is ±2 Hz (see ITU-R Recommendation BS.450-2), the tolerance on the frequency of the subcarrier during stereo broadcasts is ±6 Hz.

During monophonic broadcasts, the frequency of the subcarrier will be 57 kHz ± 6 Hz.

A.2 Subcarrier Phase

During stereo broadcasts, the subcarrier will be locked either in phase or in quadrature to the third harmonic of the 19 kHz pilot-tone. The tolerance on this phase angle is ±10°.

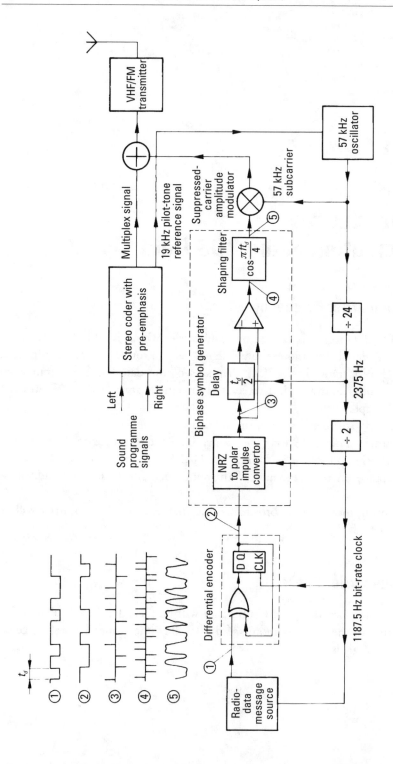

Figure A.1 Block diagram of radio-data equipment at the transmitter. (*Source:* EBU.)

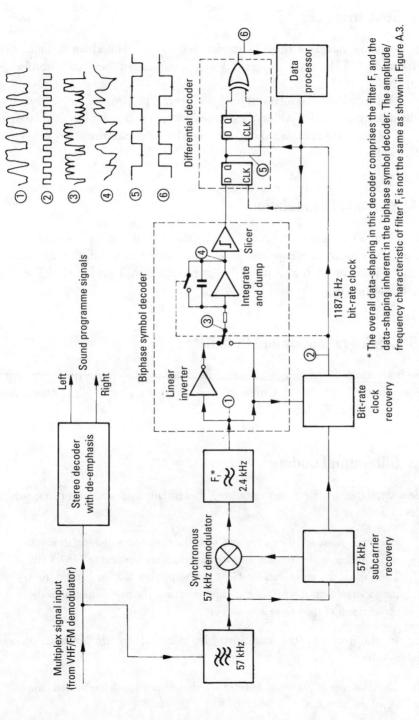

Figure A.2 Block diagram of a typical radio-data receiver/decoder. (*Source*: EBU.)

A.3 Subcarrier Level

The deviation range of the FM carrier due to the unmodulated subcarrier is from ±1.0 kHz to ±7.5 kHz. The recommended best compromise is ±2.0 kHz.

The decoder/demodulator should also operate properly when the deviation of the subcarrier is varied within these limits during periods not less than 10 ms.

The maximum permitted deviation due to the composite multiplex signal is ±75 kHz.

A.4 Method of Modulation

The subcarrier is amplitude-modulated by the shaped and biphase coded data signal. The subcarrier is suppressed. This method of modulation may alternatively be thought of as a form of two-phase phase shift keying (PSK) with a phase deviation of ±90°.

A.5 Clock-Frequency and Data Rate

The basic clock frequency is obtained by dividing the transmitted subcarrier frequency of 57 kHz by 48. Consequently, the basic data rate of the system (see Figure A.1) is 1187.5 bit/s ± 0.125 bit/s.

A.6 Differential Coding

The source data at the transmitter are differentially encoded according to the rules in Table A.1:

> where t_i is some arbitrary time and t_{i-1} is the time one message data clock period earlier, and where the message data clock rate is equal to 1187.5 Hz. Thus, when the input data level is 0, the output remains unchanged from the previous output bit. When an input 1 occurs, the new output bit is the complement of the previous output bit.

In the receiver, the data may be decoded by the inverse process (Table A.2):

> The data is thus correctly decoded whether or not the demodulated data signal is inverted.

Table A.1
Encoding Rules

Previous Output (at time t_{i-1})	New Input (at time t_i)	New Output (at time t_i)
0	0	0
0	1	1
1	0	1
1	1	0

Table A.2
Decoding Rules

Previous Input (at time t_{i-1})	New Input (at time t_i)	New Output (at time t_i)
0	0	0
0	1	1
1	0	1
1	1	0

A.7 Data Channel Spectrum Shaping

The power of the data signal at and close to the 57 kHz subcarrier is minimized by coding each source data bit as a biphase symbol. This is done to avoid data-modulated cross-talk in phase locked loop stereo decoders, and to achieve compatibility with the ARI system. The principle of the process of generation of the shaped biphase symbols is shown schematically in Figure A.1. In concept, each source bit gives rise to an odd impulse-pair, $e(t)$, such that a logic 1 at source gives:

$$e(t) = \delta(t) - \delta(t - t_d/2) \qquad (A.1)$$

and a logic 0 at source gives:

$$e(t) = -\delta(t) + \delta(t - t_d/2) \qquad (A.2)$$

These impulse pairs are then shaped by a filter $H_T(f)$, to give the required band-limited spectrum where:

$$H_T(f) = \begin{cases} \cos \dfrac{\pi f t_d}{4} & \text{if } 0 \le f \le 2/t_d \\ 0 & \text{if } f > 2/t_d \end{cases} \qquad (A.3)$$

$$t_d = \frac{1}{1187.5} \text{ s}$$

The data-spectrum shaping filtering has been split equally between the transmitter and receiver (to give optimum performance in the presence of random noise) so that, ideally, the data filtering at the receiver should be identical to that of the transmitter, i.e., as given above in Equation A.3. The overall data-channel spectrum shaping $H_0(f)$ would then be 100% cosine roll-off. The specified transmitter and receiver low-pass filter responses, as defined in Equation A.3 are illustrated in Figure A.3, and the overall data-channel spectrum shaping is shown in Figure A.4.

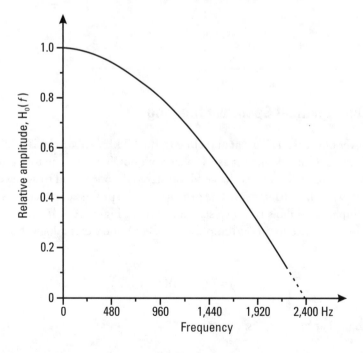

Figure A.3 Amplitude response of the specified transmitter or receiver data-shaping filter. (*Source:* EBU.)

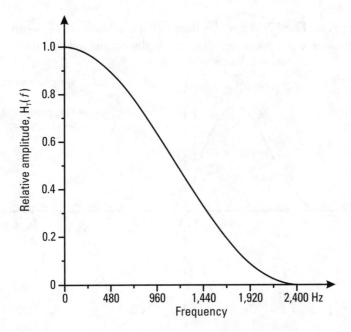

Figure A.4 Amplitude response of the combined transmitter and receiver data-shaping filters. (*Source:* EBU.)

The spectrum of the transmitted biphase coded radio data signal is shown in Figure A.5 and the time function of a single biphase symbol (as transmitted)

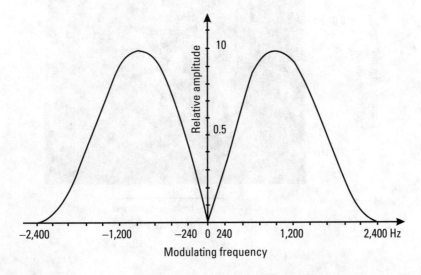

Figure A.5 Spectrum of biphase coded radio data signals. (*Source:* EBU.)

in Figure A.6. The 57 kHz radio-data signal waveform at the output of the radio-data source equipment may be seen in the photograph of Figure A.7.

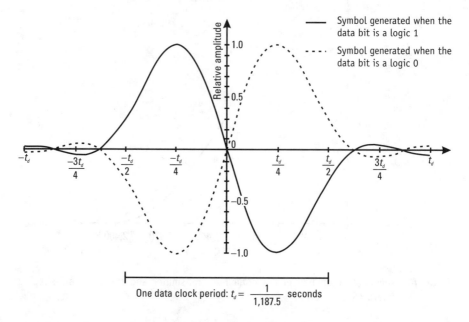

Figure A.6 Time function of a single biphase symbol. (*Source:* EBU.)

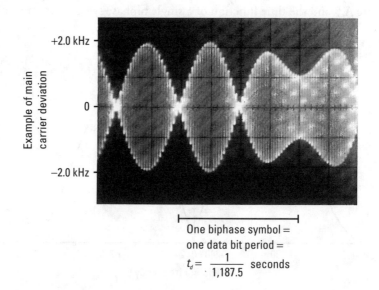

Figure A.7 57 kHz radio data signals. (*Source:* EBU.)

References

[1] CENELEC EN 50067:1998, "Specifications of the Radio Data System (RDS) for VHF/FM broadcasting, *European Committee for Electrical Standardization (CENELEC)*, 35B Rue de Stassart, B-1050, Brussels, April 1988.

Appendix B
RDS Data Decoding

B.1 Introduction

One form of a biphase symbol decoder was shown in Figure A.2 of Appendix A. In that section, it was also explained that the biphase symbol generator produces two impulses for each input bit. The resultant impulse train is the cosine-shaped filtered one, and then modulated on the 57 kHz subcarrier, the basic data rate being 1187.5 bps (i.e., 57 kHz divided by 48, which is the clock frequency). The RDS demodulator/decoder techniques, as well as their functionality and corresponding integrated circuits, were all explained in Chapter 12. In the following section, we deal entirely with decoding the baseband data stream.

B.2 Baseband Coding Structure

Figure B.1 shows the structure of the baseband coding. The largest element in the structure is called a "group" of 104 bits each. Each group is comprised of 4 blocks of 26 bits each. Each block is comprised of an information word and a check word. Each information word is comprised of 16 bits. Each check word is comprised of 10 bits.

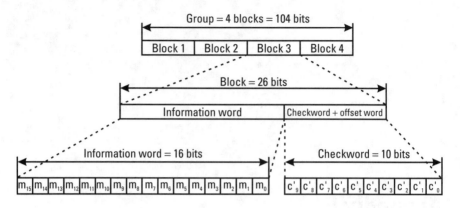

Figure B.1 Structure of the baseband coding. (*Source:* EBU.)

B.3 Order of Bit Transmission, Error Protection, and Synchronisation Information

All information words, check words, binary numbers, or binary address values have their most significant bit (msb) transmitted first (see Figure B.2). Thus, the last bit transmitted in a binary number or address has weight 2^0.

The data transmission is fully synchronous, and there are no gaps between the groups or blocks.

Check word + offset word *N*: 10 bits are added to provide error protection and block and group synchronisation information. Block 1 of any particular group is transmitted first and block 4 last. Each transmitted 26-bit block contains a 10-bit check word that is primarily intended to enable the receiver/decoder to detect and correct errors that occur in transmission. This check word (i.e., c'_9, c'_8,... c'_0 in Figure B.1) is the sum (modulo 2) of the following:

- The remainder after multiplication by x^{10} and then division (modulo 2) by the generator polynomial $g(x)$ of the 16-bit information word;
- A 10-bit binary string $d(x)$, called the "offset word," where the values are different for each block within a group (see Figure B.2 and Table B.1).

The above is true where the generator polynomial, $g(x)$ is given by

$$g(x) = x^{10} + x^8 + x^7 + x^5 + x^4 + x^3 + 1$$

Table B.1
Binary Values of the RDS Offset Words

Offset Word	Binary Value									
	d_9	d_8	d_7	d_6	d_5	d_4	d_3	d_2	d_1	d_0
A	0	0	1	1	1	1	1	1	0	0
B	0	1	1	0	0	1	1	0	0	0
C	0	1	0	1	1	0	1	0	0	0
C'	1	1	0	1	0	1	0	0	0	0
D	0	1	1	0	1	1	0	1	0	0
E	0	0	0	0	0	0	0	0	0	0

The offset words are chosen in such a way that the content in the offset register will not be interpreted as a burst of errors equal to or shorter than five bits when rotated in the polynomial shift register.

Only eight bits (i.e., d_9 to d_2) are used for identifying the offset words. The remaining two bits (i.e., d_1 and d_0) are set to logical level zero.

The five offset words A, B, C, C', and D of Table B.1 are used for all applications, except that offset word E is used in multiples of four blocks when RDS and MMBS are simultaneously implemented in the United States. Offset word E must not be used in RDS implementations outside the United States, but receivers should generally be designed to recognise offset word E and not be desynchronised when it occurs in multiples of 4.

The offset words are added (modulo 2) to the check word $c_9 - c_0$ to generate the modified check-bits: $c'_9 - c'_0$.

The purpose of adding the offset word is to provide a group and block synchronisation system in the receiver/decoder. Because the addition of the offset is reversible in the decoder, the normal additive error-correcting and detecting properties of the basic CRC code are unaffected.

The check word thus generated is transmitted msb (i.e., the coefficient of c'_9 in the check word) first and is transmitted at the end of the block that it protects.

The error-protecting code has the following error-checking capabilities:

- Detects all single and double bit errors in a block;
- Detects any single error burst spanning 10 bits or less;

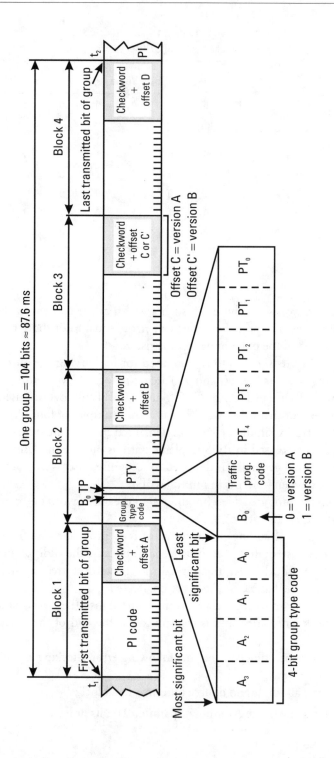

Figure B.2 Message format and addressing. (*Source:* EBU.)

- Detects about 99.8% of bursts spanning 11 bits and about 99.9% of all longer bursts.

The code is also an optimal burst error-correcting code and is capable of correcting any single burst of a span of five bits or less.

The blocks within each group are identified by the offset words A, B, C (or C′ instead of C), and D added to blocks 1, 2, 3, and 4, respectively, in each type A (or, with offset C′, type B) group.

The beginnings and ends of the data blocks may be recognised in the receiver decoder by using the fact that the error-checking decoder will, with a high level of confidence, detect block synchronisation slip as well as additive errors. This system of block synchronisation is made reliable by the addition of the offset words (which also serve to identify the blocks within the group). These offset words destroy the cyclic property of the basic code so that in the modified code, cyclic shifts of code words do not give rise to other code words.

B.4 Message Format and Addressing of Groups

The basic design principles underlying the message format and addressing structure are as follows:

- The messages that are to be repeated most frequently, and for which a short acquisition time is required (i.e., PI, PTY and TP codes, as well as the group type code) occupy the same fixed positions within every group. They can therefore be decoded without reference to any *block* outside the one that contains the information.

- There is no fixed rhythm of repetition of the various types of group; that is, there is ample flexibility to interleave the various kinds of messages to suit the needs of the transmission operator at any given time, and to allow for future developments.

- This requires addressing to identify the information content of those blocks that are not dedicated to the high-repetition-rate information.

- Each group is, so far as possible, fully addressed to identify the information content of the various blocks.

- The mixture of different kinds of messages within any one group is minimised; for example, one group type is reserved for basic tuning information, another for RadioText, and so on. This is important so that broadcasters who do not wish to transmit certain kinds of messages are not forced to waste channel capacity by transmitting groups with unused blocks. Instead, they are able to repeat more frequently those group types that contain the messages they want to transmit.

- To allow for future applications, the data formatting has been made flexible. For example, the ODA feature (see Chapter 9) may be used for the definition of any (public domain or proprietary) future application.

The main features of the message structure have been illustrated in Figure B.2. These may be seen to be as follows:

1. The first block in every group always contains a PI code.

2. The first four bits of the second block of every group are allocated to a four-bit code that specifies the application of the group (group type number address). Groups will be referred to as type number 0 to 15 according to the binary weighting $A_3 = 8$, $A_2 = 4$, $A_1 = 2$, $A_0 = 1$ (see Figure B.2). For each type (0 to 15), one of two versions, A or B, can be defined. The version is specified by the fifth bit (B_0) of block 2 as follows:

   a. $B_0 = 0$ (type A versions; e.g., 0A, 1A, etc.): the PI code is inserted in block 1 only.

   b. $B_0 = 1$ (type B versions; e.g., 0B, 1B, etc.): the PI code is inserted in block 1 and block 3 of all group types.

In general, any mixture of type A and B groups is transmitted.

3. The PTY and TP identification occupy fixed locations in block 2 of every group.

As already stated above, the PI, PTY, and TP codes can be decoded without reference to any block outside the one that contains the information. This is essential to minimise acquisition time for these kinds of messages and to retain the advantages of the short (26-bit) block length. To permit this to be done for the PI codes in block 3 of version B groups, a special offset word (which is called C′) is used in block 3 of version B groups. The occurrence of offset C′ in block 3 of any group can then be used to indicate directly that block 3 is a PI code, without any reference to the value of B_0 in block 2.

References

[1] CENELEC EN 50067:1998-Specifications of the Radio Data System (RDS) for VHF/FM broadcasting, European Committee for Electrical Standardisation (CENELEC), 35B rue de Stassart, B-1050 Brussels, April 1998.

[2] Parnall, S., "Decoding RDS, Electronics & Wireless World," Part 1, Feb. 1989, pp. 148–152, Part 2, March 1989, pp. 284–287.

[3] Ohsmann, M., "Radio Data System (RDS) Decoder," *Elektor Electronics*, Feb. 1991, pp. 56–61.

Appendix C
RDS Reception Reliability

C.1 Introduction

Field tests carried out by a number of broadcasters' research laboratories all came to the same conclusions [1–3].

Relatively low RDS data injection levels, say ±1 kHz deviation, offer a reliable data system, but only under receiving conditions with little or no multipath effects (e.g., in open, relatively flat country with good line of sight from the transmitter). Multipath effects occur due to reflections of the transmitted signal by high-rise buildings or in mountainous terrain. In a moving receiver, once multipath effects occur, there is a sharp decrease in reception reliability. However, reliability can be increased by data repetition because multipath reception has the effect of reducing reception performance for fractions of a second as the receiver moves from good to poor to good reception locations.

C.2 Bit Error Rate

Curve (a) in Figure C.1 shows the bit error rate of the experimental reference RDS demodulator measured as a function of the power applied at the aerial input of the VHF/FM receiver. The VHF/FM receiver used in these measurements had a noise figure of 5.5 dB. For the purposes of comparison, the theoretical bit error rate expected according to theory is given as curve (b). It may be

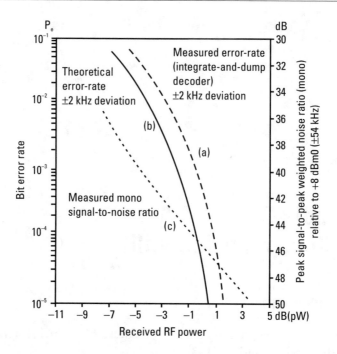

Figure C.1 Bit error rate of the reference RDS demodulator (theoretical and measured results): receiver noise figure = 5.5 dB. (*Source:* BBC/EBU.)

seen that the measured performance of this experimental RDS decoder is within about 1 dB of that expected from theory.

Note that the deviation due to the RDS signal was ±2 kHz. For other deviations, the results may be linearly scaled. For example, for ±1.2 kHz deviation, the RF input power needed to attain a given bit error rate would be about 4.4 dB (i.e., 20 log (2/1.2)) greater.

Curves (a) and (b) in Figure C.1 illustrate the rapid failure with the typical declining signal-to-noise ratio characteristic of most digital systems. For satisfactory operation, the RDS system needs a bit error rate better than about 2 in 10^2.

Also shown in Figure C.1, as curve (c), is the mono peak signal-to-peak weighted noise ratio (measured according to ITU-R Recommendation BS.468-4) obtained in the sound programme channel of the same VHF/FM receiver used in the error rate measurements of curve (a). This was measured after 50 μs de-emphasis relative to a +8 dBm tone at 440 Hz (i.e., 54 kHz deviation with standard BBC line-up levels). Stereo reception would, in theory at least, need 20 dB more RF power to achieve the same signal-to-noise ratio as that shown for mono.

Thus, it is found that when the only impairment to reception is random noise due to low field-strength, the RDS system operates satisfactorily until beyond the point at which stereo reception of the programme signal becomes unusable. At the field-strength corresponding to the failure point of the RDS system, mono reception is noisy but still intelligible and remains so for aerial input levels down to about −10 dB(pW). In practice, this is not a problem because such low field-strengths usually occur only well outside the service area of the transmitter, and it is often possible to switch to an alternative frequency carrying the same service. Furthermore, where such low field-strengths prevail, there are usually other impairments to reception, such as multipath, which render the programme signal unusable.

C.3 Block Error Rate

It is important to remember that because of the use of differential decoding in the RDS demodulator, the errors usually occur in bursts spanning two bits. Single errors occur only when adjacent bits in the received data stream, before differential decoding, are in error.

In the RDS system, the block length for the purpose of error protection is 26 bits. Curves showing the probability of correct reception of PI codes (which are equivalent to one RDS block) are given in Figure C.2, where the only impairment to reception is random noise due to low field strength and is plotted as a function of RF input power to the aerial input of the VHF/FM receiver. Curves (4) and (3) show the results with and without error correction, respectively, for ±2 kHz deviation; curves (2) and (1) show the corresponding results for ±1.2 kHz deviation. The error correction used in obtaining the results shown in curves (4) and (2) accorded with that recommended in the RDS/RBDS standards; that is, bursts of errors spanning only up to two bits in a block were corrected—longer bursts were detected and those blocks were rejected.

Comparing curves (3) and (4) with curves (2) and (1), it may be seen that under conditions where the received signal is impaired by low field-strength only, the use of error correction yields a maximum improvement equivalent to less than 2 dB in RF level. However, as will be shown, error correction yields a greater improvement with the burst errors characteristic of mobile reception.

C.4 Error Rates for RDS Messages

As was noted above, the probability of correct reception of a PI code is simply the probability of reception of an RDS block, as shown by the curves in

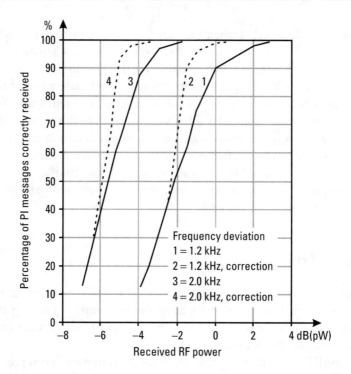

Figure C.2 Influence of the RF level, error correction, and frequency deviation on the broadcast reliability. (*Source:* IRT/EBU.)

Figure C.2. The same is also true for all other kinds of messages that occupy fixed positions within *all* RDS group types and can therefore be decoded without reference to any information outside the block that contains them. Included in this category are PI, PTY, and TP identification.

Other messages require correct decoding of the group type address, and perhaps other information such as the segment address for PS codes. Inevitably, the more RDS blocks that a message and its related addressing occupy, the lower the probability of correct reception of the complete message. However, in most applications needing long messages (e.g., RadioText), it is not necessary to receive the whole message correctly before making use of it; missing or erroneous characters can easily be tolerated on displayed messages.

Curves showing the probability of correct reception of PI, PS, and RT messages when the only impairment to reception is random noise due to low field-strength are given in Figure C.3. The deviation due to the RDS signal was ±2 kHz, and two-bit burst error correction was applied.

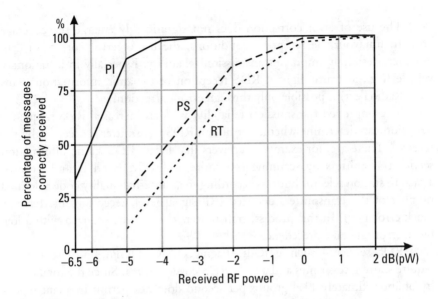

Figure C.3 Reliability of reception of RDS messages. (*Source:* IRT/EBU.)

C.5 RDS Coverage Area

In a study carried out by the IRT in Germany [3] for PI (corresponding to block error rate) and PS features, and for various reception conditions, the relation between RDS frequency deviation and coverage area was determined using frequency deviations of ±1.2 kHz, ±1.5 kHz, and ±2.0 kHz.

The total of all areas with transmission reliability equal to or greater than 90% was designated as the RDS coverage area in this study. To avoid time-consuming field measurements, the experiments were performed with a radio channel simulator.

The study came to the following conclusions.

In unfavourable reception conditions, transmission reliability and thus coverage area can be increased by raising the RDS frequency deviation. The increase in coverage area is directly dependent on the reception conditions prevailing in that area. In unfavourable reception conditions, the coverage area can be distinctly enlarged with an increase in RDS frequency deviation higher than ±1.5 kHz.

Simultaneous transmission of ARI reduces the coverage area. Services requiring relatively brief access times should therefore not be installed on ARI networks.

The use of error correction does not significantly enlarge the coverage area. In unfavourable reception conditions, the use of error correction can, however, definitely improve transmission reliability. Especially in a stationary vehicle (e.g., at traffic lights), RDS reception of dynamic information items (e.g., RadioText) is possible only through error correction.

The improved transmission reliability with an increased RDS frequency deviation can determine whether or not an RDS transmitter is accepted by the receiver. There are, for example, receivers that locate RDS traffic information service transmitters by scanning the entire VHF/FM Band II and store or attach to station selector buttons according to an internal quality code a defined number of the transmitters detected (RF signal level, reception interference, block error rate). In this process, transmitters that can be received with a low block error rate take precedence.

RDS services—such as Radio Paging, TMC, DGPS, and also EWS—require short access times and a large RDS coverage area. Since the total capacity of approximately 11.4 groups per second does not permit frequent repetition of these dynamic information items, these requirements can be met only through a higher RDS frequency deviation and the resulting improvement in transmission reliability.

References

[1] Mielke, J., and K. H. Schwaiger, "Progress with the RDS system and experimental results," *EBU Review-Technical*, No. 217, June 1986.

[2] Lyner, A. G., Experimental Radio Data System (RDS): A survey of reception reliability in the UK, BBC Research Department, Report No. 1987/17.

[3] Schwaiger K. H., The RDS coverage area in relation to the RDS frequency deviation, Technical Report B 148/95, Institut für Rundfunktechnik München, 1995.

Appendix D
Required Data Repetition Rates for Programme-Related RDS Features

The mixture of different kinds of messages within any one group is minimised; for example, one group type is reserved for basic tuning information, another for RadioText, and so on. This is important so that broadcasters that do not wish to transmit messages of certain kinds are not forced to waste channel capacity by transmitting groups with unused blocks. Instead, they are able to repeat more frequently those group types that contain the messages they want to transmit. There is no fixed rhythm of repetition of the various types of group; that is, there is ample flexibility to interleave the various kinds of messages to suit the needs of the users at any given time and to allow for future developments.

A total of four type 0A groups are required to transmit the entire PS name and therefore four type 0A groups will be required per second. The repetition rate of the type 0A group may be reduced if more capacity is needed for other applications (see Table D.1). However, with EON receivers, search tuning is affected by the repetition rate of type 0 groups. A minimum of two type 0A groups per second is necessary to ensure correct functioning of PS and AF features. It should be noted that in this case, transmission of the complete PS will take two seconds. However, under typical reception conditions, the introduction of errors will cause the receiver to take four seconds or more to acquire the PS name for display.

Table D.1

Recommended Repetition Rates for the Programme Related Features

Main Features	Group Types That Contain This Information	Appropriate Repetition Rate Per Second
Programme Identification (PI) code	All	11.4 [1]
Programme Type (PTY) code	All	11.4 [1]
Traffic Programme (TP) identification code	All	11.4 [1]
Programme Service (PS) name [4]	0A, 0B	1
Alternative Frequency (AF) code pairs	0A	4
Traffic Announcement (TA) code	0A, 0B, 15B	4
Decoder Identification (DI) code	0A, 0B, 15B	1
Music Speech (MS) code	0A, 0B, 15B	4
RadioText (RT) message	2A, 2B	0.2 [2]
Enhanced Other Networks (EON) information	14A, 14B	Up to 2 [3]

1) Valid codes for this item will normally be transmitted with at least this repetition rate whenever the transmitter carries a normal broadcast programme.

2) A total of 16 type 2A groups are required to transmit a 64-character RT message and therefore 3.2 type 2A groups will be required per second.

3) The maximum cycle time for the transmission of *all* data relating to *all* cross-referenced PSs shall be less than two minutes.

4) PS must only be used for identifying the PS, and it must not be used for other messages giving sequential information.

When the PIN or slow-labelling is implemented, type 1 groups should be used once per minute, but when the PIN is changed, a type 1 group should be repeated four times with a separation of about 0.5 seconds. Where RP is implemented in RDS, a type 1A group will be transmitted in an invariable sequence, regularly once per second, except at each full minute, where it is replaced by one type 4A group.

The mixture of groups shown in Table D.2 is suitable to meet the repetition rates noted above.

Table D.2
Group Repetition Rates

Group Types	Features	Typical Proportion of Groups of This Type Transmitted
0A or 0B	PI, PS, PTY, TP, AF [1], TA, DI, M/S	40%
1A or 1B	PI, PTY, TP, PIN	10%
2A or 2B	PI, PTY, TP, RT	15% [2]
14A or 14B	PI, PTY, TP, EON	10%
Any other	Other applications	25%

1) Type 0A group only.

2) Assuming that type 2A groups are used to transmit a 32-character RadioText message. A mixture of type 2A and 2B groups in any given message should be avoided.

References

[1]   CENELEC EN 50067:1998 - Specifications of the Radio Data System (RDS) for VHF/FM broadcasting, European Committee for Electrical Standardisation (CENELEC), 35B rue de Stassart, B-1050 Brussels, April 1998.

Appendix E
RDS Data Transmission Capacity Limits

E.1 Introduction

The use of the features to achieve automated tuning, essentially for mobile reception, requires a considerable part of the channel capacity. An RDS uses a data stream of 1187.5 bps and offers a choice of some 20 well-defined features. The data capacity is not large enough to allow all these features to be implemented in the same channel, as shown in the following analysis.

E.2 Calculation of RDS Capacity

The bit rate of 1187.5 bps of the RDS is fairly low and therefore the RDS channel capacity is a rather limited resource. Bearing in mind that the four check words in each group of 104 bits occupy a total of 40 bits, and that each group address needs 5 bits, one arrives at a *useful* bit rate of

$$1187.5 - (1187.5/104) \times (40 + 5) = 673.7 \text{ bps}$$

E.3 Analysis of RDS Capacity

E.3.1 Grouping of Different Features

In order to analyse the usage of RDS capacity for each of the features, they are grouped together in categories according to their impact on the RDS channel capacity. Firstly, the capacity required for all the broadcasting programme-related features is examined and then what is left over for the nonprogramme-related features can be identified.

As is shown in Table E.1, the programme-related features can be divided into four categories:

- The five primary features—AF, PI, PS, and TP/TA—mainly required for the automated tuning process;
- The EON information, which really is a complementary feature for automated tuning as far as networks are concerned;
- A group of features—CT, DI, MS, PIN, PTY, and PTYI—requiring relatively little RDS capacity to be implemented. Note that PIN only requires a repetition rate of one group type 1A per minute; however, in connection with RP it must be increased to one per second.
- RadioText.

Furthermore, in Table E.1 the nonprogramme-related features—IH, RP, TDC, and TMC—are identified.

E 3.2 Discussion on the Analysis

The following can be seen from Table E.1:

- *AF, PI, PS, TP/TA:* These primary functions of RDS, essentially supporting the automated tuning process, require already 48.35% of the available channel capacity.
- *EON:* If the automated tuning process were to be enhanced by broadcasters operating networks through the use of the EON feature, 8.24% of the capacity would have to be set aside, bringing the total up to 56.59%.
- *PTY, MS, DI, PIN, CT:* All these features require relatively little RDS capacity. While the first three require 9.65% whether implemented or not, for reasons of coding, the latter two add an additional 0.21%

Table E.1
Analysis of the Required RDS Capacity

Application	Feature	Group Types Containing this Information	Appropriate Minimum Group Repetition Rate per Second	Number of Occupied Bits per Group	Number of Occupied Bits per Second	Percent of 673.7 Bits per Second	Accumulated RDS Capacity
Automated tuning	PI	All	11.4	16	182.4	27.07	
	PS	0A	4.0	16	64.0	9.50	
	AF	0A	4.0	16	64.0	9.50	
	TP	all	11.4	1	11.4	1.69	
	TA	0A	4.0	1	4.0	0.59	48.35
Enhanced information on other networks	EON	14A(B)	1.5	37	55.5	8.24	56.59
Various other programme-related features	PTY	All	11.4	5	57.0	8.46	
	MS	0A	4.0	1	4.0	0.59	
	DI	0A	4.0	1	4.0	0.59	
	PIN	1A(B)	0.02/1.0* 37**	0.74/37*	0.11/5.5*		
	CT	4A	0.02	34	0.68	0.10	66.44/71.83*

Table E.1 (continued)

Application	Feature	Group Types Containing this Information	Appropriate Minimum Group Repetition Rate per Second	Number of Occupied Bits per Group	Number of Occupied Bits per Second	Percent of 673.7 Bits per Second	Accumulated RDS Capacity
RadioText	RT	2A(B)	3.2***	37	118.4	17.58	84/89.41*
Various nonprogramme-related features	RP	1A/4A/7A/13A					
	TDC	5A(B)					
	IH	6A(B)					
	TMC	3A/8A					

* If Radio Paging is used, an increased repetition rate of 1 per second is necessary.

** Although 16 bits are actually used for PIN, the associated 16 undefined and 5 spare bits must also be taken into account since this represents used capacity.

*** To agree with Note 2 in Table 4 of CENELEC EN 50 067:1998, Section 3.1.3: A total of 16 type 2A groups are required to transmit a 64 character RadioText message, and therefore to transmit this message in 5 seconds, 3.2 type 2A groups will be required per second.

The appropriate minimum repetition rate is specified in the RDS and RBDS standards and is detailed in Appendix D herein.

(5.6% if RP is implemented), bringing the total required to 66.44% (or 71.83% with RP).

- *RT*: RadioText requires an additional 17.58%, bringing the total required for the programme-related features up to 84% (or 89.41% with RP).

This then leaves 16% (10% with RP), for the implementation of the nonprogramme-related features (IH, ODA, TDC, and TMC); that is, 108 bps (or 68 bps with RP). Each of them will require, if implemented, a significant proportion of the remainder. In fact, in the case of Radio Paging, the paging traffic is likely to account for the balance of the capacity remaining.

Type A groups, for example, can carry in ODA only up to 37 application bits, which means that there will only be capacity left for two to three groups per second. This makes the dilemma of the very restricted data capacity in RDS for other than radio programme-related applications very obvious.

However, due to the time-multiplexing possibilities of many of these features, the average capacity available for other features will be greater than this indicated "peak demand," since it is unlikely, for example, that continuous RadioText will be a requirement.

E.4 Conclusion About RDS Capacity

This form of analysis makes it quite clear that, from the broadcasters' point of view, it seems advisable to set aside about 80% of the RDS channel capacity for the implementation of all programme-related features. This leaves about one-fifth of the channel capacity to accommodate features unrelated to the radiated programmes. This implies that only a limited number of further applications can actually be implemented within any specific programme service. Due consideration must therefore be given to how IH, ODA, RP, TDC, and TMC can be distributed among different programme networks; for example, a network that carries TP/TA could only accommodate in addition, say, TMC. The same is also true with respect to so far unknown applications (e.g., those that can be implemented through ODA); their required capacity has to be carefully examined before any decision can be taken as to the possibility of their implementation.

In conclusion, RDS has a finite capacity—it is not an inexhaustible resource. That is a fundamental reality that it shares with any other communication channel, including, of course, the VHF/FM broadcast channels that RDS exploits and serves.

References

[1]   CENELEC EN 50067:1998 - Specifications of the Radio Data System (RDS) for VHF/FM broadcasting European Committee for Electrical Standardisation (CENELEC), 35B rue de Stassart, B-1050 Brussels, April 1998.

[2]   EBU, Guidelines for the implementation of the RDS system, Tech. 3260-E, Jan. 1990.

Appendix F
PI Coding in RDS and RBDS

F.1 Introduction

PI codes are the key to the automatic tuning and follow-me features of a RDS. They act as "digital signatures" to prevent false switching between stations carrying different programme signals (but perhaps sharing alternate frequencies). Care must be taken, therefore, in the region of borders between different countries, to avoid two stations having the same PI code that are both receivable at the same location.

Codes shall be assigned in each country in such a way that automatic search tuning to other transmitters radiating the same programme can locate the same PI code; that is, all 16 bits shall be identical. In cases where during a few programme hours, a network is split to radiate different programmes, each of these programmes shall carry a different PI code, by using different coverage-area codes.

PI codes in North America are issued and utilised differently than in the rest of the world. The particular principles used are explained in Section F.3.

F.2 Programme Identification Code Structure

The PI code consists of 16 bits and is usually referred to by four hexadecimal characters, or nibbles. See Figure F.1 for an example.

b_{15}			b_{12}	b_{11}			b_8	b_7			b_4	b_3			b_0

Bits b_{15} to b_{12} (first nibble): Country identification (Codes are given in Appendix G)
Bits b_{11} to b_8 (second nibble): Programme TYpe in terms of area coverage
Bits b_7 to b_0 (third and fourth nibbles): Programme reference number
General remark: All codes are binary-coded hex numbers

Figure F.1 PI code structure. (*Source:* EBU.)

Code assignments for bits b_{11} to b_0 should be decided by relevant authorities in each country individually.

F.2.1 Country Identification

Extended Country Codes shall be transmitted in type 1A groups to render the country code in bits b_{15} to b_{12} of the PI code *unique*. The ECC is carried in variant 0 of block 3 of type 1A groups and consists of eight bits. This variant should be transmitted at least once every minute.

The bit allocation of the Extended Country Codes is given in Figure F.2, and the codes are given in Appendix G.

F.2.2 Coverage Area Codes

The second nibble is allocated according to the coverage area of the service. Code 0 hex indicates a local service broadcast from a single transmitter only,

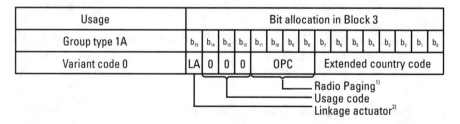

1. The operator code for Radio Paging using the Enhanced Paging Protocol is defined in Appendix M of the RDS/RBDS standards.
2. The linkage actuator is defined in the method for linking RDS programme services (see the RDS/RBDS standards).

Figure F.2 Structure of variant 0 of block 3 of type 1A groups (Extended Country Codes). (*Source:* EBU.)

code 1 indicates a service available internationally, code 2 hex, a national service, and code 3 hex, a service covering a national region or other major area within a country code area. Codes 4-F hex are used for regional services.

Hex-coding rules for bits b_{11} to b_8 are given in Figure F.3.

F.2.3 Programme Reference Number

Hex coding for bits b_7 to b_0 are given in Table F.1.

Example: Decimal number 148 corresponds to hexadecimal-code 94—this is just a straight dec to hex conversion.

In order to clearly identify the different programme families, these codes should, in each country, be systematically assigned and generically linked to the programme families.

Area coverage code	L	I	N	S	R1	R2	R3	R4	R5	R6	R7	R8	R9	R10	R11	R12
HEX	0	1	2	3	4	5	6	7	8	9	A	B	C	D	E	F

I = International	The same programme is also transmitted in other countries
N = National	The same programme is transmitted throughout the country
S = Supra-regional	The same programme is transmitted throughout a large part of the country
R1 to R12 = Regional	The programme is available only in one location or region over one or more frequencies, and there exists no definition of its frontiers
L = Local	Local programme transmitted via a single transmitter (only) during the whole transmission time

Figure F.3 Coverage area hex codes used in the second nibble (bits b_{11} to b_8) of the PI code. (*Source:* EBU.)

Table F.1
Bits b_7 to b_0

Decimal Numbers	Hex	
00	00	Not assigned
01 to 255	01 to FF	Programme reference number

F.3 PI Coding Rules for North America

F.3.1 Basic Principles

In North America (United States, Canada, and Mexico), PI codes are based on call letters rather than being assigned by any national organisation, as is done in countries throughout the rest of the world. A portion of the PI codes are reserved for network usage and also for assignment to stations in Canada and Mexico.

The CENELEC RDS standard accepts the usage of coverage area codes for all possible PI codes (see Section F.2 above on coverage area codes). Within North America coverage, area codes are only recognised in the following range of PI codes:

- B_01 to B_FF hex

- D_01 to D_FF hex

- E_01 to E_FF hex

All other PI codes do not make use of coverage area codes and must be handled as such within the receiver.

Some current European receivers store PI codes into presets in addition to storing frequencies into presets. This function is to recognise the broadcast first by programme rather than frequency. Thus, if a preset is pushed and the PI code has changed, the European RDS receivers would not recognise the new PI code and go into a PI search.

A technical publication from the EBU states:

> If, however, the PI code changes completely, the receiver should initiate a PI search for a frequency whose PI code exactly matches the PI code of the original tuned frequency. Failing an exact PI code match, the receiver should search for a PI code differing only in the regional element (b_{11} to b_8) from the original PI code. If neither of these criterion are met, the receiver should remain on the original tuned frequency.

Therefore, since call letters are used to create the PI code, the receiver would have to do a PI search every time a station would change call letters or a preset is pushed in a new listening area having a station at the same frequency as the preset station. For PI codes smaller than B000 hex, future receivers could check the AF list associated with a preset and if no AFs are acceptable, a PI search could be initiated. If no identical PI is found, the receiver should return to the original tuned frequency and accept the new PI code.

If a PI search is performed, the regional variant search (the second search to match PI codes, differing only in b_{11} to b_8) should be eliminated in a PI search if the tuned PI is below B000 hex, or within the ranges of C001 to CFFF hex and F001 to FFFF hex.

If a feature similar to European regional variants is desired, a grouping in the B, D, and E blocks could be designated as follows. If NPR broadcasts break off national programming to go local for a period of time, it could be assigned a PI of B_01. Note that you cannot use 0 as the second nibble because current receivers will not search for AFs. Therefore, use 1–8 to F for indication of a "regional" variant. If no AFs or identical PIs are found via the AF list or an identical PI search, the receiver could, while tuned to NPR station 1 (PI = B101) accept a variant NPR station 2 whose PI varies only in the second nibble (bits 5–8). Thus B201, B301, B401, ... could be accepted.

PI codes starting with the B, D, and E nibbles yield 765 possibilities for "regional" programming. These PI codes will be shared by the United States, Canada, and Mexico. The problem here becomes that a "telephone book" needs to be kept; however, there should not be too many broadcasts that fit in this category and not many would be used.

F.3.2 Call Letter Conversion Method

The call letter conversion method to be used is described in detail in the RBDS standard, Annex D, Section 6.1. It makes uses of a conversion table and some simple algorithms. The details required to make the conversion are, however, so extensive that they cannot be quoted in this book, and only the RBDS standard itself can help here in proceeding with the conversion of a particular case.

The rules that are spelled out in the RBDS standard could also be implemented in encoder software so that the station that wants to use RDS can simply enter its call letters and subsequently the encoder would then make the PI code assignment in full compliance with these rules. The broadcast station interested in this matter should discuss this issue with the encoder supplier to avoid any error or misunderstanding.

References

[1] CENELEC EN 50067:1998 - Specifications of the Radio Data System (RDS.) for VHF/FM broadcasting European Committee for Electrical Standardisation (CENELEC), 35B rue de Stassart, B-1050 Brussels, April 1998.

[2] EIA/NAB(1997), National Radio Systems Committee: United States RBDS Standard version 2.0, Specification of the Radio Broadcast Data System (RBDS).

[3] EBU, Guidelines for the implementation of the RDS system, Tech. 3260-E, Jan. 1990.

Appendix G
RDS Country or Area Identification Codes

G.1 Introduction

This appendix lists the RDS country or area identification codes as they are allocated in the RDS/RBDS standards [1, 2]. Although these standards were developed for Europe and North America only, the continued development of RDS worldwide made it necessary to allocate in the RDS/RBDS standards PI country/area identification symbols throughout the world.

The PI code is used in each RDS group, always in the first block of type A and B groups; for type B groups, the PI code is repeated in the third block. Details of the PI code usage are given in Chapter 3 and Appendix F.

The country/area code is a hex code symbol that occupies the first four bits of the PI code. Only 15 different symbols exist, and they are repetitively assigned to countries in such a way that broadcasts of the one country/area sharing the same symbol cannot be received in the other country/area. However, a unique identification of any country/area sharing the country/area symbol with other countries/areas is possible if the optional ECC has also been implemented.

It should also be noted that certain countries use more than one country/area symbol. Germany, for example, has two symbols for historical reasons, since the respective codes were allocated before the reunification. Before, one code was for Eastern Germany and the other for Western Germany, but now no distinction is being made and both codes may be used all over Germany. In the United States, Canada, and Mexico, a range of country/area codes is used to permit application of a formula for the automatic conversion of call letters into the PI code to avoid the need for any centralised PI administration in these countries.

G.2 PI Code Structure

The Programme Identification code consists of 16 bits. Bits b_{15} to b_{12} make up the country/area symbol (hex code) as defined in Section 4. Code assignments b_{11} to b_0 should be assigned by relevant authorities in each country or area in such a way, for example, that bits b_{11} to b_8 describe the Programme TYpe in terms of area coverage, and bits b_7 to b_0 are a programme reference number.

Codes shall always be assigned in such a way that automatic search tuning to other transmitters radiating the same programme can locate the same Programme Identification Code; that is, all 16 bits shall be identical.

G.3 Extended Country Codes

ECC shall be transmitted in type 1A groups (the last eight bits in Block 3 of Variant 0) to render the country/area code in bits 1 to 4 of the PI code unique.

G.4 Allocated Country/Area Symbols

G.4.1 Allocation of Symbols for Countries in ITU Region 1

G.4.1.1 European Broadcasting Area

Table G.1 lists the PI and ECC symbols for European countries.

Table G.1
European Codes

Country/Area	ISO Code	Symbol for PI (Hex)	ECC (Hex)
Albania	AL	9	E0
Algeria	DZ	2	E0
Andorra	AD	3	E0

Table G.1 (continued)

Country/Area	ISO Code	Symbol for PI (Hex)	ECC (Hex)
Austria	AT	A	E0
Azores [Portugal]	PT	8	E4
Belgium	BE	6	E0
Belarus	BY	F	E3
Bosnia-Herzegovina	BA	F	E4
Bulgaria	BG	8	E1
Canaries [Spain]	ES	E	E2
Croatia	HR	C	E3
Cyprus	CY	2	E1
Czech Republic	CZ	2	E2
Denmark	DK	9	E1
Egypt	EG	F	E0
Estonia	EE	2	E4
Faroe Islands [Denmark]	DK	9	E1
Finland	FI	6	E1
France	FR	F	E1
Germany	DE	D or 1	E0
Gibraltar [United Kingdom]	GI	A	E1
Greece	GR	1	E1
Hungary	HU	B	E0
Iceland	IS	A	E2
Iraq	IQ	B	E1
Ireland	IE	2	E3
Israel	IL	4	E0
Italy	IT	5	E0
Jordan	JO	5	E1
Latvia	LV	9	E3
Lebanon	LB	A	E3
Libya	LY	D	E1
Liechtenstein	LI	9	E2
Lithuania	LT	C	E2
Luxembourg	LU	7	E1

Table G.1 (continued)

Country/Area	ISO Code	Symbol for PI (Hex)	ECC (Hex)
Macedonia	MK	4	E3
Madeira [Portugal]	PT	8	E4
Malta	MT	C	E0
Moldova	MD	1	E4
Monaco	MC	B	E2
Morocco	MA	1	E2
Netherlands	NL	8	E3
Norway	NO	F	E2
Palestine	PS	8	E0
Poland	PL	3	E2
Portugal	PT	8	E4
Romania	RO	E	E1
Russian Federation	RU	7	E0
San Marino	SM	3	E1
Slovakia	SK	5	E2
Slovenia	SI	9	E4
Spain	ES	E	E2
Sweden	SE	E	E3
Switzerland	CH	4	E1
Syrian Arab Republic	SY	6	E2
Tunisia	TN	7	E2
Turkey	TR	3	E3
Ukraine	UA	6	E4
United Kingdom	GB	C	E1
Vatican City State	VA	4	E2
Yugoslavia	YU	D	E2

G.4.1.2 African Broadcasting Area

Table G.2 lists the PI and ECC symbols for African countries.

Table G.2
African Codes

Country/Area	ISO Code	Symbol for PI (Hex)	ECC (Hex)
Ascension Island	??	A	D1
Cabinda	??	4	D3
Angola	AO	6	D0
Algeria	DZ	2	E0
Burundi	BI	9	D1
Benin	BJ	E	D0
Burkina Faso	BF	B	D0
Botswana	BW	B	D1
Cameroon	CM	1	D0
Canary Islands [Spain]	ES	E	E0
Central African Republic	CF	2	D0
Chad	TD	9	D2
Congo	CG	C	D0
Comoros	KM	C	D1
Cape Verde	CV	6	D1
Cote d'Ivoire	CI	C	D2
Democratic Rep. of Congo	ZR	B	D2
Djibouti	DJ	3	D0
Egypt	EG	F	E0
Ethiopia	ET	E	D1
Gabon	GA	8	D0
Ghana	GH	3	D1
Gambia	GM	8	D1
Guinea-Bissau	GW	A	D2
Equatorial Guinea	GQ	7	D0
Republic of Guinea	GN	9	D0
Kenya	KE	6	D2
Liberia	LR	2	D1
Libya	LY	D	E1
Lesotho	LS	6	D3
Mauritius	MU	A	D3

Table G.2 (continued)

Country/Area	ISO Code	Symbol for PI (Hex)	ECC (Hex)
Madagascar	MG	4	D0
Mali	ML	5	D0
Mozambique	MZ	3	D2
Morocco	MA	1	E2
Mauritania	MR	4	D1
Malawi	MW	F	D0
Niger	NE	8	D2
Nigeria	NG	F	D1
Namibia	NA	·1	D1
Rwanda	RW	5	D3
Sao Tome & Principe	ST	5	D1
Seychelles	SC	8	D3
Senegal	SN	7	D1
Sierra Leone	SL	1	D2
Somalia	SO	7	D2
South Africa	ZA	A	D0
Sudan	SD	C	D3
Swaziland	SZ	5	D2
Togo	TG	D	D0
Tunisia	TN	7	E2
Tanzania	TZ	D	D1
Uganda	UG	4	D2
Western Sahara	EH	3	D3
Zambia	ZM	E	D2
Zanzibar	??	D	D2
Zimbabwe	ZW	2	D2

Note: Countries or areas with no known ISO code are shown with a ?? code.

G.4.1.3 Former Soviet Union

Table G.3 lists the PI and ECC symbols for Soviet Union countries.

Table G.3
Soviet Union Codes

Country/Area	ISO Code	Symbol for PI (Hex)	ECC (Hex)
Armenia	AM	A	E4
Azerbaijan	AZ	B	E3
Belarus	BY	F	E3
Estonia	EE	2	E4
Georgia	GE	C	E4
Kazakhstan	KZ	D	E3
Kyrgyzstan	KG	3	E4
Latvia	LV	9	E3
Lithuania	LT	C	E2
Moldova	MD	1	E4
Russian Federation	RU	7	E0
Tajikistan	TJ	5	E3
Turkmenistan	TM	E	E4
Ukraine	UA	6	E4
Uzbekistan	UZ	B	E4

G.4.2 Allocations of Symbols for Countries in ITU Region 2

Table G.4 lists the PI and ECC symbols for countries in ITU Region 2.

Table G.4
ITU Region 2 Codes

Country/Area	ISO Code	Symbol for PI (Hex)	ECC (Hex)
Anguilla	AI	1	A2
Antigua and Barbuda	AG	2	A2
Argentina	AR	A	A2
Aruba	AW	3	A4
Bahamas	BS	F	A2

Table G.4 (continued)

Country/Area	ISO Code	Symbol for PI (Hex)	ECC (Hex)
Barbados	BB	5	A2
Belize	BZ	6	A2
Bermuda	BM	C	A2
Bolivia	BO	1	A3
Brazil	BR	B	A2
Canada	CA	B, C, D, E	A1
Cayman Islands	KY	7	A2
Chile	CL	C	A3
Colombia	CO	2	A3
Costa Rica	CR	8	A2
Cuba	CU	9	A2
Dominica	DM	A	A3
Dominican Republic	DO	B	A3
Ecuador	EC	3	A2
El Salvador	SV	C	A4
Falkland Islands	FK	4	A2
Greenland	GL	F	A1
Grenada	GD	D	A3
Guadeloupe	GP	E	A2
Guatemala	GT	1	A4
Guiana	GF	5	A3
Guyana	GY	F	A3
Haiti	HT	D	A4
Honduras	HN	2	A4
Jamaica	JM	3	A3
Martinique	MQ	4	A3
Mexico	MX	B, D, E, F	A5
Montserrat	MS	5	A4
Netherlands Antilles	AN	D	A2
Nicaragua	NI	7	A3
Panama	A	9	A3
Paraguay	PY	6	A3

Table G.4 (continued)

Country/Area	ISO Code	Symbol for PI (Hex)	ECC (Hex)
Peru	PE	7	A4
Puerto Rico	PR	1..9, A, B, D, E	A0
Saint Kitts	KN	A	A4
Saint Lucia	LC	B	A4
St Pierre and Miquelon	PM	F	A6
Saint Vincent	VC	C	A5
Suriname	SR	8	A4
Trinidad and Tobago	TT	6	A4
Turks and Caicos Islands	TC	E	A3
United States of America	US	1..9, A, B, D, E	A0
Uruguay	UY	9	A4
Venezuela	VE	E	A4
Virgin Islands [British]	VG	F	A5
Virgin Islands [USA]	VI	1..9, A, B, D, E	A0

G.4.3 Allocations of Symbols for Countries in ITU Region 3

Table G.5 lists the PI and ECC symbols for countries in ITU Region 3.

Table G.5
ITU Region 3 Codes

Country/Area	ISO Code	Symbol for PI (Hex)	ECC (Hex)
Afghanistan	AF	A	F0
Saudi Arabia	SA	9	F0
Australia	AU		
Australian Capital Territory		1	F0
New South Wales		2	F0
Victoria		3	F0
Queensland		4	F0
South Australia		5	F0

Table G.5 (continued)

Country/Area	ISO Code	Symbol for PI (Hex)	ECC (Hex)
Australia (continued)	AU		
Western Australia		6	F0
Tasmania		7	F0
Northern Territory		8	F0
Bangladesh	BD	3	F1
Bahrain	BH	E	F0
Myanmar [Burma]	MM	B	F0
Brunei Darussalam	BN	B	F1
Bhutan	BT	2	F1
Cambodia	KH	3	F2
China	CN	C	F0
Sri Lanka	LK	C	F1
Fiji	FJ	5	F1
Hong Kong	HK	F	F1
India	IN	5	F2
Indonesia	ID	C	F2
Iran	IR	8	F0
Iraq	IQ	B	E1
Japan	JP	9	F2
Kiribati	KI	1	F1
Korea [South]	KR	E	F1
Korea [North]	KP	D	F0
Kuwait	KW	1	F2
Laos	LA	1	F3
Macau	MO	6	F2
Malaysia	MY	F	F0
Maldives	MV	B	F2
Micronesia	FM	E	F3
Mongolia	MN	F	F3
Nepal	NP	E	F2
Nauru	NR	7	F1
New Zealand	NZ	9	F1

Table G.5 (continued)

Country/Area	ISO Code	Symbol for PI (Hex)	ECC (Hex)
Oman	OM	6	F1
Pakistan	PK	4	F1
Philippines	PH	8	F2
Papua New Guinea	PG	9	F3
Qatar	QA	2	F2
Solomon Islands	SB	A	F1
Western Samoa	WS	4	F2
Singapore	SG	A	F2
Taiwan	TW	D	F1
Thailand	TH	2	F3
Tonga	TO	3	F3
UAE	AE	D	F2
Vietnam	VN	7	F2
Vanuatu	VU	F	F2
Yemen	YE	B	F3

References

[1]   CENELEC EN 50067:1998 - Specifications of the Radio Data System (RDS) for VHF/FM broadcasting European Committee for Electrical Standardisation (CENELEC), 35B rue de Stassart, B-1050 Brussels, April 1998.

[2]   EIA/NAB, National Radio Systems Committee: United States RBDS Standard - Specification of the Radio Broadcast Data System (RBDS), Washington D.C., April 1998.

Appendix H
PTY Display Terms in Several Different Languages

H.1 Introduction

This Appendix gives details for the following languages: English, French, German, Spanish and Swedish.

Each of the following tables have a listing of the PTY codes from EN50067:1998 Annex F, showing the number, English language Programme TYpe and then a corresponding term (and not necessarily the literally translated one) of the other appropriate language, followed by their 8-character and 16-character display terms which may be used.

The listings in the different languages given here, were agreed upon within the RDS Forum. Corresponding terms for other languages can be found, when agreed, on the RDS Forum Web site, URL: www.rds.org.uk/.

It is very important to note, that "LARGE LETTERS" (upper case) and "small letters" (lower case) are used and that "spaces" are clearly shown using the "_" notation. Not all receivers can display both "LARGE LETTERS" and "small letters," but they are specified, so that receivers which do display them can benefit from the enhanced readability of their use.

Table H.1

PTY Code Descriptions and Abbreviations in Multiple Languages: English

	English	English	English
Number	Programme type	8-character display	16-character display
0	No programme type or undefined	None	None
1	News	News	News
2	Current Affairs	Affairs	Current _ Affairs
3	Information	Info	Information
4	Sport	Sport	Sport
5	Education	Educate	Education
6	Drama	Drama	Drama
7	Culture	Culture	Cultures
8	Science	Science	Science
9	Varied	Varied	Varied _ Speech
10	Pop Music	Pop _ M	Pop _ Music
11	Rock Music	Rock _ M	Rock _ Music
12	Easy Listening Music	Easy _ M	Easy _ Listening
13	Light Classical Music	Light _ M	Light _ Classics _ M
14	Serious Classical Music	Classics	Serious _ Classics
15	Other Music	Other _ M	Other _ Music
16	Weather	Weather	Weather _ & _ Metr
17	Finance	Finance	Finance
18	Children's Programmes	Children	Children's _ Progs
19	Social Affairs	Social	Social _ Affairs
20	Religion	Religion	Religion
21	Phone In	Phone _ In	Phone _ In
22	Travel	Travel	Travel _ & _ Touring
23	Leisure	Leisure	Leisure & Hobby
24	Jazz Music	Jazz	Jazz _ Music
25	Country Music	Country	Country _ Music
26	National Music	Nation _ M	National _ Music
27	Oldies Music	Oldies	Oldies _ Music
28	Folk Music	Folk _ M	Folk _ Music
29	Documentary	Document	Documentary
30	Alarm Test	TEST	Alarm _ Test
31	Alarm	Alarm !	Alarm - Alarm !

Note: Large letters, small letters and the spaces are critical.

Table H.2
PTY Code Descriptions and Abbreviations in Multiple Languages: French

	English	French	French	French
Number	Programme type	Programme type	8-character display	16-character display
0	No programme type or undefined	Codification de programme non utilisée ou genre non défini	Aucun	Non défini
1	News	Informations	Infos	Informations
2	Current Affairs	Magazine	Magazine	Magazine
3	Information	Info-Service	Services	Info-Service
4	Sport	Sport	Sport	Sport
5	Education	Education	Educatif	Educatif
6	Drama	Dramatique	Fiction	Fiction
7	Culture	Culture	Culture	Culture
8	Science	Sciences	Sciences	Sciences
9	Varied	Divertissement	Divers	Divertissement
10	Pop Music	Pop	M _ Pop	Musique _ Pop
11	Rock Music	Rock	M _ Rock	Musique _ Rock
12	Easy Listening Music	Chansons	Chansons	Chansons
13	Light Classical Music	Classique Léger	M _ Cl _ Lég	Classique _ Léger
14	Serious Classical Music	Musique Classique	Classiq	MusiqueClassique
15	Other Music	Autre Musique	Autre _ M	Autre _ Musique
16	Weather	Météo	Météo	Météo
17	Finance	Economie et Finances	Economie	Economie
18	Children's Programmes	Programmes pour enfants	Enfants	Enfants
19	Social Affairs	Société	Société	Société
20	Religion	Religion	Religion	Religion
21	Phone In	Ligne ouverte et interactivité	Forum	Ligne _ ouverte
22	Travel	Voyages	Voyages	Voyages
23	Leisure	Loisirs	Loisirs	Loisirs
24	Jazz Music	Musique de jazz	Jazz	Jazz
25	Country Music	Musique de country	Country	Country
26	National Music	Chansons du pays	Ch _ pays	Chanson _ du _ pays
27	Oldies Music	Musique rétro	Rétro	Musique _ rétro
28	Folk Music	Musique folklorique	Folklore	Folklore
29	Documentary	Documentaire	Document	Documentaire
30	Alarm Test	Test d'alarme	TEST	TEST
31	Alarm	Alerte	Alerte _ !	Alerte _ !

Note: Large letters, small letters and the spaces are critical.

Table H.3

PTY Code Descriptions and Abbreviations in Multiple Languages: German

	English	German	German	German
Number	Programme type	Programme type	8-character display	16-character display
0	No programme type or undefined	Keine Programmartkennung	Kein _ PTY	Kein _ PTY
1	News	Nachrichtendienst	Nachrich	Nachrichten
2	Current Affairs	Politik und Zeitgeschehen	Aktuell	Aktuelle _ Info
3	Information	Service Programm	Service	Service _ Programm
4	Sport	Sport	Sport	Sport
5	Education	Lernen und Weiterbildung	Bildung	Bildung
6	Drama	Hörspiel und Literatur	Hör+Lit	Hörspiel _ + _ Lit
7	Culture	Kultur, Kirche und Gesellschaft	Kultur	Kultur+Gesellsch
8	Science	Wissenschaft	Wissen	Wissenschaft
9	Varied	Unterhaltendes Wort	Unterh	Unterhaltung
10	Pop Music	Popmusik	Pop	Pop _ Musik
11	Rock Music	Rockmusik	Rock	Rock _ Musik
12	Easy Listening Music	Unterhaltungsmusik	U-Musik	Unterhalt _ Musik
13	Light Classical Music	Leichte klassische Musik	L-Musik	Leichte _ Klassik
14	Serious Classical Music	Ernste klassische Musik	E-Musik	Ernste _ Klassik
15	Other Music	Spezielle Musikprogramme	--Musik	Spezielle _ Musik
16	Weather	Wetter	Wetter	Wetter
17	Finance	Wirtschaft	Wirtsch	Wirtschaft
18	Children's Programmes	Kinderprogramm	Kinder	Kinderprogramm
19	Social Affairs	Soziales	Soziales	Soziales
20	Religion	Religion	Religion	Religion
21	Phone In	Anrufsendung	Anruf	Anrufsendung
22	Travel	Reiseinformation	Reise	Reiseinformation
23	Leisure	Freizeit	Freizeit	Freizeit
24	Jazz Music	Jazz	Jazz	Jazz
25	Country Music	Countrymusik	Country	Country _ Musik
26	National Music	Musik des Landes	Landes _ M	Landesmusik
27	Oldies Music	Oldiemusik	Oldies	Oldies _ Musik
28	Folk Music	Folklore	Folklore	Folklore
29	Documentary	Feature	Feature	Feature
30	Alarm Test	Alarmtest	Test	Alarm _ Test
31	Alarm	Alarm	Alarm!	Alarm! _ Alarm!

Note: Large letters, small letters and the spaces are critical.

Table H.4
PTY Code Descriptions and Abbreviations in Multiple Languages: Spanish

	English	Spanish	Spanish	Spanish
Number	Programme type	Programme type	8-character display	16-character display
0	No programme type or undefined	No se indica el tipo programa	Ninguno	Ninguno
1	News	Noticias	Noticias	Noticias
2	Current Affairs	Magazin	Magazine	Magazine
3	Information	Información	Info	Información
4	Sport	Deportes	Deportes	Deportes
5	Education	Educación	Educa	Educación
6	Drama	Drama	Drama	Drama
7	Culture	Cultura	Cultura	Cultura
8	Science	Ciencia	Ciencia	Ciencia
9	Varied	Varios	Varios	Varios
10	Pop Music	Música Pop	M _ Pop	Música _ Pop
11	Rock Music	Música Rock	M _ Rock	Música _ Rock
12	Easy Listening Music	Grand éxitos	G _ éxitos	Grandes _ éxitos
13	Light Classical Music	Música Ligera	M _ Ligera	Música _ Ligera
14	Serious Classical Music	Música Clásica	Clásica	Música _ Clásica
15	Other Music	Otra Música	M _ Varia	Música _ Varia
16	Weather	Información Metereológica	Meteo	Meteorología
17	Finance	Economia	Economía	Economía
18	Children's Programmes	Infancia	Infancia	Infancia
19	Social Affairs	Sociedad	Sociedad	Sociedad
20	Religion	Religión	Religión	Religión
21	Phone In	Opinión	Opinión	Opinión-Oyentes
22	Travel	Viajes	Viajes	Viajes
23	Leisure	Ocio	Ocio	Ocio
24	Jazz Music	Música Jazz	M _ Jazz	Música _ Jazz
25	Country Music	Música Country	Country	Música _ Country
26	National Music	Música Nacional	Nacional	Música _ Nacional
27	Oldies Music	Música de ayer	M _ Ayer	Música _ de _ Ayer
28	Folk Music	Música Folk	M _ Folk	Música _ Folk
29	Documentary	Documental	Docu	Documental
30	Alarm Test	Test de Alarma	Prueba	Prueba _ de _ Alarma
31	Alarm	Alarma	Alarma	¡ _ Alarma _ !

Note: Large letters, small letters and the spaces are critical.

Table H.5
PTY Code Descriptions and Abbreviations in Multiple Languages: Swedish

	English	Swedish	Swedish	Swedish
Number	Programme type	Programme type	8-character display	16-character display
0	No programme type or undefined	Programtypen saknas eller är odefinierad	Pty _ sakn	Pty _ saknas
1	News	Korta nyheter	Nyheter	Nyheter
2	Current Affairs	Fördjupning av nyheter	Aktuellt	Aktualiteter
3	Information	Allmän information	Info	Information
4	Sport	Sport	Sport	Sport
5	Education	Utbildningsprogram	Utbildn	Utbildning
6	Drama	Teater och program om teater	Teater	Teater
7	Culture	Kultur i bred mening	Kultur	Kultur
8	Science	Vetenskapsprogram	Vetenskp	Vetenskap
9	Varied	Underhållningsprogram	Underh	Underhållning
10	Pop Music	Populärmusik	Pop	Popmusik
11	Rock Music	Rockmusik	Rock	Rockmusik
12	Easy Listening Music	Underhållningsmusik	Lättlyss	Lättlyssnat
13	Light Classical Music	Lätt klassisk musik(ej kompletta verk)	L _ klass	Lätt _ klassiskt
14	Serious Classical Music	Kompletta klassiska verk	Klassisk	Klassisk _ musik
15	Other Music	Övrig Musik	Övrig _ m	Övrig _ musik
16	Weather	Väderraporter	Väder	Väder
17	Finance	Ekonomiprogram	Ekonomi	Ekonomi
18	Children's Programmes	Barnprogram	För _ barn	För _ barn
19	Social Affairs	Program om sociala frågor	Socialt	Sociala _ frågor
20	Religion	Andliga frågor	Andligt	Andliga _ frågor
21	Phone In	Telefonväkteri	Telefon	Telefonväkteri
22	Travel	Resor och semester	Resor	Resor_&_semester
23	Leisure	Fritid och hobby	Fritid	Fritid _ & _ hobby
24	Jazz Music	Jazzmusik	Jazz	Jazzmusik
25	Country Music	Countrymusik	Country	Countrymusik
26	National Music	Nationell musik	Nation _ m	Nationell _ musik
27	Oldies Music	Klassisk pop	Oldies	Gamla _ godingar
28	Folk Music	Folkmusik	Folkm	Folkmusik
29	Documentary	Dokumentärer	Dokument	Dokumentärer
30	Alarm Test	Provlarm	Prov	Prov
31	Alarm	Larm	!!Larm!!	!!Larm!!

Note: Large letters, small letters and the spaces are critical.

References

[1]   PTY display terms in several different languages, RDS Forum Document 98/009, Sept. 1998.

Appendix I
Character Sets for Alphanumeric Display

Three different alphanumeric character repertoires have been defined in Annex E of the RDS/RBDS standards; they are reproduced in Figures E.1 to E.3. Taken together, they permit the composition of texts indicating the name of the programme service and the constitution of radio data messages (e.g., RadioText) or alphanumeric paging calls. The three code tables were designed by the EBU. They contain almost all the characters in the international reference version of ISO Publication 646[1]. The same codes have been given to each of these characters in all three tables. Care has been taken in the design of the coding tables to ensure that it will be possible to satisfy all the requirements within large geographical areas with each repertoire, and it is therefore likely that some receivers will be equipped to display only the characters included in one of the three repertoires. Up until 1997/98, RDS receivers have used only Figure I.1, or even only parts of it. Nonetheless, it is necessary to provide information identifying the repertoire in use in order to ensure that the display corresponds as closely as possible to the intentions of the broadcasting organisation when received on a receiver able to display characters from more than one

---

1. Including the figures 0 to 9 and punctuation; nonetheless, in certain cases, codes have been re-allocated to characters taken from the EBU repertoires, in accordance with the provision of ISO Publication 646.

repertoire. Since the default code table is Figure I.1, this indication is only required if one of the other two code tables is used or if one changes from one of them back to code table 1.

The repertoire tables were designed by the EBU with the view of covering the requirements satisfying the use of languages within the European broadcasting area. However, a compromise had to be made to keep these tables small in size. As a consequence of this, one or the other character from a particular language was left out, because it could be replaced by another. For example, in Greek, small theta, should be substituted by a capital theta.

In accordance with the practice in the videotext service, where more than one character repertoire is defined also, control codes have been allocated to distinguish between the basic (Figure I.1) and two auxiliary (see Annex E of the RDS/RBDS standards) code tables. The selection of the required code table is controlled by the transmission of the corresponding repertoire control characters—SI (0/15), SO (0/14), and LS2 (1/11 followed by 6/14)[2]. In radio data, it is controlled by the transmission of one of the following pairs of repertoire control characters:

- 0/15, 0/15—code table of Figure I.1;

- 0/14, 0/14—code table of Figure E.2 in Annex E of the RDS/RBDS standards;

- 1/11, 6/14—code table of Figure E.3 in Annex E of the RDS/RBDS standards.

These characters do not occupy a space in the display, but have effect on the displayable characters having the *same address*, and on all characters having numerically higher addresses up to but not including the address of another repertoire control character. In default of a repertoire control character, the display coding taking effect at address 0 should be assumed to be in accordance with the default code table given in Figure I.1. For example, the name "Radio 1" could be transmitted in type 0 groups as follows:

Characters:	Ra	di	oSP	1SP
Text segment address:	0	1	2	3

---

2. The notation A/B is used to designate the character appearing in column A on line B in the table.

The text segment address: 0, 1, 2, 3 uses the character codes in pairs, respectively:

5/2,	6/1;
6/4,	6/9;
6/15,	2/0;
3/1,	2/0

to convey the name "Radio 1".

		Displayable characters from the code table of ISO Norm 646						Additional displayable characters for:							
								EBU common-core (7 languages)		Complete Latin-based repertoire (25 languages)					
b7		0	0	0	0	0	0	1	1	1	1	1	1	1	1
b6		0	0	1	1	1	1	0	0	·0	0	1	1	1	1
b5		1	1	0	0	1	1	0	0	1	1	0	0	1	1
b4		0	1	0	1	0	1	0	1	0	1	0	1	0	1
b3 b2 b1 b0		2	3	4	5	6	7	8	9	10	11	12	13	14	15
0 0 0 0	0		0	@	P	‖	p	ā	à	ª	º	Á	Å	Ã	ã
0 0 0 1	1	!	1	A	Q	a	q	à	ä	α	'	À	Ä	Å	å
0 0 1 0	2	"	2	B	R	b	r	é	ê	©	²	É	Ê	Æ	æ
0 0 1 1	3	#	3	C	S	c	s	è	ë	‰	³	È	Ë	Œ	œ
0 1 0 0	4	¤	4	D	T	d	t	í	î	Ğ	±	Í	Î	ŷ	ŵ
0 1 0 1	5	%	5	E	U	e	u	ì	ï	ě	İ	Ì	Ï	Ý	ý
0 1 1 0	6	&	6	F	V	f	v	ó	ô	ń	ń	Ó	Ô	Õ	õ
0 1 1 1	7	'	7	G	W	g	w	ò	ö	ő	ű	Ò	Ö	Ø	ø
1 0 0 0	8	(	8	H	X	h	x	ú	û	π	μ	Ú	Û	Þ	þ
1 0 0 1	9	)	9	I	Y	i	y	ù	ü	Œ	¿	Ù	Ü	Ŋ	ŋ
1 0 1 0	10	*	:	J	Z	j	z	Ñ	ñ	£	÷	Ř	ř	Ŕ	ŕ
1 0 1 1	11	+	;	K	[(1)	k	{ (1)	Ç	ç	$	°	Č	č	Ć	ć
1 1 0 0	12	,	<	L	\	l	|	Ş	ş	←	¼	Š	š	Ś	ś
1 1 0 1	13	-	=	M	] (1)	m	} (1)	β	ğ	↑	½	Ž	ž	Ź	ź
1 1 1 0	14	.	>	N		n		ı	ı	→	¾	Ð	đ	Ŧ	ŧ
1 1 1 1	15	/	?	O		o		IJ	ij	↓	§	Ŀ	ŀ	ß	

Attention is drawn to the fact that low cost receivers may be able to display only the characters in Column 2 lines 0, 7, 12, 13, 14 and 15; Column 3 lines 0 to 9; Column 4 lines 1 to 15; Column 5 lines 0 to 10.

Figure I.1 Basic default code table for 218 displayable characters forming the complete EBU Latin-based repertoire. The characters shown in positions marked (¹) in the table are those of the "international reference version" of ISO Publication 646 that do not appear in the complete Latin-based repertoire given in Appendix 2 of EBU document Tech. 3232 (2nd edition, 1982). (*Source:* EBU.)

References

[1]   CENELEC EN 50067:1998 - Specifications of the Radio Data System (RDS) for VHF/FM broadcasting, European Committee for Electrical Standardisation (CENELEC), 35B rue de Stassart, B-1050 Brussels, April 1998.

[2]   EIA/NAB(1997) - National Radio Systems Committee: United States RBDS Standard Version 2.0, Specification of the Radio Broadcast Data System (RBDS).

Appendix J
Implemented RDS Features in Various Countries

J.1 Introduction

When broadcasters or transmission operators implement RDS, it is relatively easy to implement the static RDS features such as PI, PS, and AF, because a stand-alone RDS encoder at a transmitter site can be used. But programme-related RDS features require a data link from the studio to the RDS encoder, which may be inexpensive for just one single transmitter but quite expensive when the radio programme is going over a network of transmitters. An additional complexity exists when the network is split during certain hours of the day for regionalised programming. This is one of the main reasons why some major broadcast networks have not yet fully implemented the dynamic programme-related RDS features. Another reason is, of course, related to the fact that the production of radio programme-dependent RDS data requires additional human resources, and there is still some reluctance to invest in this activity given the fact that the number of receivers that support the dynamic RDS programme-related features (and specifically RadioText) is still relatively small in relation to the total number of receivers that are in use. This situation is, however, slowly changing because new RDS receivers tend to increasingly support dynamic programme-related RDS features.

The RDS Forum has for many years published a list of RDS features implemented in various countries. The following provides a summary that is derived from the data published in that list.

Almost all operators use a subcarrier injection level of ±2 kHz. When this level is different, this fact is mentioned below.

Austria: ORF uses, in addition to the static RDS features, TP/TA and EON. RT is only tested, and not really implemented. CT is not used. PTY and traffic message channel TMC are being considered.

Belgium: VRT and RTBF use in addition to the static RDS features also TP/TA and EON. Only VRT uses PTY and RT. All these programmes use CT. TMC is being considered, and RTBF will start operation in late 1998.

Brazil: Interest is increasing and initial services are starting on several radio stations in Rio de Janeiro and São Paulo.

Croatia: Croatian radio uses in addition to the static RDS features also TP/TA and EON. CT is implemented and so is RP using an injection level of ±4 kHz.

Denmark: DR uses, in addition to the static RDS features, TP/TA and EON. CT is implemented, but not PTY. RadioText is only tested, and not really implemented. TMC and DGPS will be implemented during 1998.

Finland: YLE uses almost all RDS features except MS. However PTY, PIN, and RT are only on two of their five networks. RP is used on two networks that use the higher injection level of ±4 kHz.

France: Radio France uses, in addition to the static RDS features, TP/TA and EON. RT is used only in Paris on FIP. TMC is used in Paris and is planned for use by the motorway radio programme for 1998/99. RP is used on programmes of Radio France. The injection level in use is ±4 kHz. Several commercial radio stations use TP, but they do not switch TA. CT is generally implemented on all stations. MS, PTY, and PIN are not yet used.

Germany: The public broadcasters use almost all RDS features, but many of them are only implemented partially. CT is sometimes not correctly implemented. PIN is not used. TMC was implemented in 1997 and for DGPS the RASANT system will be used nationwide. RP is partially used. Private broadcasters use (sometimes wrongly) dynamic PS. Several programmes use programme-related RadioText.

Greece: ERT uses CT and RT in addition to the static RDS features, but not EON. Private broadcasters often wrongly use dynamic PS.

Hong Kong: RTHK uses only the static RDS features and is testing TP, TA, PTY, and RT.

Hungary: Some programmes use the static RDS features, RT, and RP with an injection level of ±4 kHz.

Ireland: RTE uses mainly static RDS features, but also PIN and EON. PTY and RT are planned. RP is used by RTE for its own staff, the injection level being ±2.5 kHz. Their other networks use a deviation of ±2.2 kHz.

Italy: RAI mainly uses the static RDS features and is testing the other ones. CT is not used. TMC is being tested in Northern Italy, starting from Bologna. Private radio stations use dynamic PS wrongly. Many of these use frequencies that are "offset" by 50 kHz.

Luxembourg: RTL uses mainly the static RDS features and CT. DGPS is being tested.

Netherlands: Radio programmes use the static RDS features and CT. RT is frequently implemented. Not used are PTY, PIN, and MS. RP and DGPS are partially used. A regular TMC service starts in the spring of 1998 on Radio 3. Some commercial radio stations wrongly use dynamic PS.

Norway: NRK programmes uses the static RDS features, EON, CT, and RT. Local radios use also RT. The DGPS is being tested.

Portugal: Most radio stations use the static RDS features. RDP plans to start TMC in Lisbon in 1998.

Slovenia: Radio Slovenia uses, in addition to the static RDS features, TP/TA and EON. CT is implemented and so is RP, using an injection level of ±4 kHz.

South Africa: SABC uses only the static RDS features and EON. RP is being tested.

Spain: RNE uses almost all RDS features. TMC will be implemented in Madrid and on the motorway to Barcelona and France. Private networks use mainly the static RDS features. PTY is used by some networks.

Sweden: SR has implemented all RDS features. Teracom operates DGPS and the Swedish Road Authority uses TMC nationwide. RP was in use for many years and has been closed down. Injection levels are either ±2.5 or 3.0 kHz. Local radio stations use mainly the static RDS features and some test RT. Some private radio stations wrongly use dynamic PS.

Switzerland: SRG/SSR mostly use the static RDS features and EON. DGPS and TMC are planned. Some private stations wrongly use dynamic PS.

United Kingdom: BBC has implemented most RDS features except MS. Classic FM has implemented the static RDS features and RT. Furthermore, DGPS is implemented under the Additional Services Licence and on the same transmitters. TMC will be implemented in 1998. PTY and CT are also used on many independent radio stations.

United States: Radio stations are increasingly interested in using RDS, and in all metropolitan areas, several local radio stations now offer RDS, usually implemented with static features. There is much interest in programme-related features, with trials of RT and PTY in progress.

Appendix K
Web Site of the RDS Forum

The Web site of the RDS Forum provides many necessary references and up-to-date information for users of the RDS technology. It also offers the possibility to join the RDS Forum. The structure of the Web site is as outlined in Figure K.1.

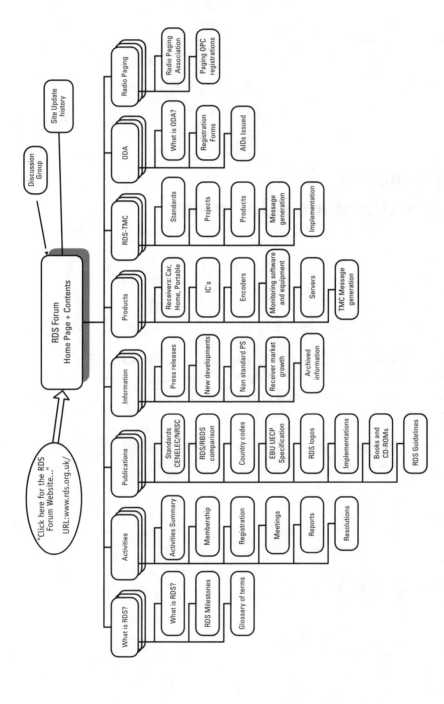

Figure K.1 The structure of the RDS Forum Web site. (*Source*: RDS Forum.)

Appendix L
UECP Message Commands

L.1 Introduction

In Chapter 11, the Universal Encoder Communication Protocol (UECP) is described. This appendix gives more detail about the command format, shows how it is applied in two specific examples, and adds a list of all possible message commands to permit the reader to get an overview of the functionality that the UECP permits in data communications with RDS encoders.

L.2 Command Format

The message description is made as shown in Figure L.1. The first column indicates the descriptor of the message, which is detailed in the second column.

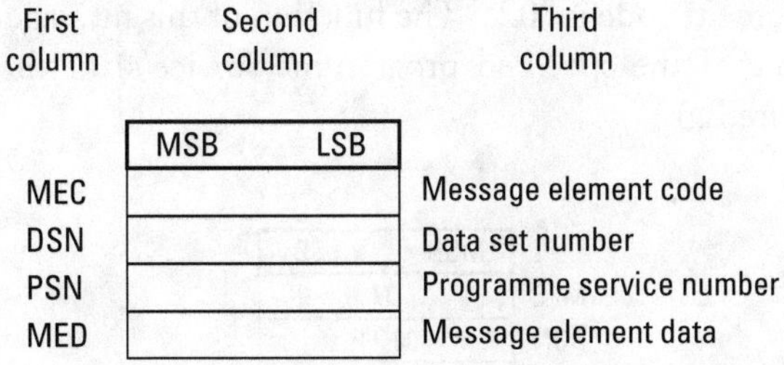

Figure L.1 UECP message command format. (*Source:* EBU.)

Each element in Figure L.1 presents one byte where the bits are numbered from 7–0 (from left to right). For transmission of a respective message, each byte is represented by two hex symbols, of which the permitted range is indicated in the respective element. The message structure used is explained in Chapter 11, Figure 11.5.

The third column in Figure L.1 gives additional information for each row and describes the nature of the data that may be entered.

The coding of all RDS features is in the same format as used in the RDS/RBDS standards and if not, usually this is specifically mentioned in the particular command.

L.3 Examples of Specific Messages

L.3.1 Message to Set the PTY Code

The message element code is "07." The function of this message command is to set the Programme TYpe information of the specified programme service on the specified data set(s). (See Figure L.2.)

Example

<07><00><05><08>

Set PTY to 8 in current data set, programme service 5.

L.3.2 Message to Set the PS Name

The message element code is "02." The function of this message command is to set the PS name of the specified programme service(s) of the specified data set(s). (See Figure L.3.)

	MSB LSB	
MEC	07	
DSN	00..FF	
PSN	00..FF	
MED	00..1F	PTY

Figure L.2 UECP message command to set the PTY code. (*Source:* EBU.)

	MSB	LSB	
MEC	02		
DSN	00..FF		
PSN	00..FF		
MED	20..FE		PS Character 1
MED	20..FE		PS Character 2
MED	20..FE		PS Character 3
MED	20..FE		PS Character 4
MED	20..FE		PS Character 5
MED	20..FE		PS Character 6
MED	20..FE		PS Character 7
MED	20..FE		PS Character 8

Figure L.3 UECP message command to set the PS name. (*Source:* EBU.)

The characters of the PS name are coded in conformity with the character sets described in Appendix I.

Example

<02><00><02><52><41><44><49><4F><20><31><20>

Set PS in current data set for programme service 2 to >RADIO_1_<

L.4 Listing of all Possible UECP Version 5.1 Message Commands

L.4.1 RDS Message Commands

1. PI;
2. PS;
3. PIN;
4. DI;
5. TA/TP;
6. MS;
7. PTY;
8. PTYN;

9. RT;

10. AF;

11. EON–AF;

12. Slow labelling codes;

13. Linkage information.

L.4.2 Open Data Application Commands

1. ODA configuration and short message command;

2. ODA identification group usage sequence;

3. ODA free-format group;

4. ODA relative priority group sequence;

5. ODA "burst mode" control;

6. ODA "spinning wheel" timing control.

L.4.3 Transparent Data Commands

1. TDC;

2. EWS;

3. IH;

4. TMC;

5. Free-format group.

L.4.4 Paging Commands

1. Paging call without message;

2. Paging call with numeric message (10 digits);

3. Paging call with numeric message (18 digits);

4. Paging call with alphanumeric message (80 characters);

5. International paging with numeric message (15 digits);

6. International paging with functions message;

7. Transmitter network group designation;

8. EPP transmitter information;

9. EPP call without additional message;

10. EPP national and international call with alphanumeric message;
11. EPP national and international call with variable-length numeric message;
12. EPP national and international call with variable-length functions message.

L.4.5   Clock Setting and Control

1. Real-time clock;
2. Real-time clock correction;
3. CT On/Off.

L.4.6   RDS Adjustment and Control

1. RDS On/Off;
2. RDS phase;
3. RDS level.

L.4.7   ARI Adjustment and Control

1. ARI On/Off;
2. ARI area;
3. ARI level.

L.4.8   Control and Setup Commands

1. Site address.
2. Encoder address;
3. Make PSN list;
4. PSN enable/disable;
5. Communication mode;
6. TA control;
7. EON-TA control;
8. Reference input select;
9. Data set select;

10. Group sequence;

11. Group variant code sequence;

12. Extended group sequence;

13. PS character code table selection;

14. Encoder access right;

15. Communications port mode configuration;

16. Communications port speed configuration;

17. Communications port timeout configuration.

L.4.9 Bidirectional Commands (Remote and Configuration Commands)

1. Message acknowledgement;

2. Request message.

L.4.10 Specific Message Commands

1. Manufacturer's specific command.

References

[1]   EBU, Universal Encoder Communications Protocol UECP-version 5.1, European Broad-casting Union, 17A, Ancienne Route, CH-1218 Geneva, Switzerland Aug. 1997.

Appendix M
Glossary and Abbreviations

M.1 Glossary

This glossary provides brief descriptions of the RDS features and some other definitions that are useful in the overall context of broadcasting. The RDS features follow, in alphabetical order.

M.1.1 Alternative Frequencies (AF) List

The AF feature is intended to give information on the frequency of the various transmitters broadcasting the *same* programme in the same or adjacent reception areas, to enable receivers equipped with a memory to store the list(s), and to reduce the time for switching to another transmitter. This facility is particularly useful for mobile car and portable receivers (see Chapter 3).

M.1.2 Clock Time (CT) and Date

The CT feature is intended to update/synchronise a free-running clock in an RDS receiver. In conformity with ITU-R Recommendations, broadcast time and date codes use coordinated universal time (UTC) and modified Julian date (MJD). However the listener/user will not use this information directly and the conversion to local time and date will be made in the receiver's circuitry. CT is used as a time stamp by various RDS applications and thus it *must* be transmitted accurately set (see Chapter 5).

M.1.3 Decoder Identification (DI) and Dynamic PTY Indicator (PTYI)

The DI and dynamic PTYI features are indicators that use designated bits to indicate which of several possible operating modes are in use with the broadcast audio, to allow a decoder to be set in a receiver to the correct mode, and to indicate if PTY codes in the transmission are dynamically switched (see Chapter 4).

M.1.4 Extended Country Code (ECC)

The ECC feature is intended to supplement information supplied by the PI that together provide a unique country code identification. The PI code only permits the definition of 15 different country codes, which are shared, so they do not permit the unique identification of the country of broadcasting origin (not necessarily of transmission). The ECC feature (using eight bits) together with the country identification in the PI code, describe a unique combination (see Chapter 3).

M.1.5 Enhanced Other Networks (EON) Information

The EON feature can be used to update information stored in a receiver about programme services *other than* the one being received. AFs, the PS name, the TP and TA flags, as well as PTY and PIN information can all be transmitted for each other service in a two minute period (see Chapter 5).

M.1.6 Emergency Warning System (EWS)

The EWS feature is intended to provide for the coding of confidential warning messages. These messages will be broadcast only in case of systems testing and actual emergency, and will only be evaluated by special receivers (see Chapter 5).

M.1.7 In-House Applications (IH)

The IH feature allows broadcasters or transmission operators to use a small amount of the RDS data capacity for their own purposes, such as monitoring equipment, systems status, and remote switching. The applications of this data may be decided by each operator and will only be evaluated by special receivers.

M.1.8 Music Speech Switch (MS)

The MS indicator feature provides a two-state flag signal to inform a receiver that either music or speech is being broadcast at that time. This signalling may permit a receiver equipped with two separate volume controls (one for music and one for speech) to be automatically selected for these two conditions so that the listener could preadjust the balance to suit individual listening requirements.

M.1.9 Open Data Applications (ODA)

The ODA feature is designed to allow data applications not previously specified in EN 50067:1998 [1] to be conveyed in a number of allocated groups in an RDS transmission. The groups allocated are indicated by the use of type 3A groups, which are used to identify to a receiver the particular data application in use in accordance with the registration details in the EBU/RDS Forum Open Data Applications Directory (see Chapter 9).

M.1.10 Programme Identification (PI)

The PI feature is a code intended to help a receiver manage received data and particularly to discern the specific programme, associated data, and area/country of the broadcast. The code is not intended for direct display and is assigned to individual radio programme services to enable it to be distinguished from all other programme services. One important application of this code is to enable a receiver to search automatically for an alternative frequency in the case of bad reception of the programme to which the receiver is tuned. The criteria for changeover to a new frequency is the presence of a better signal having the *same* PI code (see Chapter 3).

M.1.11 Programme Item Number (PIN)

The PIN feature comprises a simple code to enable receivers and recorders designed to make use of this feature to respond to a particular programme item(s) that a user has preselected. Use is made of the scheduled programme time, to which the day of the month is added in order to avoid ambiguity (see Chapter 4).

M.1.12 Programme Service (PS) Name

The PS name feature is designed to provide information for an RDS receiver to display the name of the radio programme service instead of, for example, the tuned frequency being displayed. The PS is formatted for eight character displays using either LARGE or small characters—for example, >Classic_<. The PS name is intended to inform the listener what programme service is being broadcast. The programme service name is not intended to be used for automatic search tuning and must *not* be used for giving sequential information (see Chapter 4).

M.1.13 Programme TYpe (PTY)

The PTY feature uses an identification code, transmitted with each programme item, which is intended to specify the current Programme Type from 29 possibilities. This code may be used for various tuning modes and can assist a suitable receiver and/or recorder to be preset to respond only to programme items of the desired type. The PTY feature also carries an alarm functionality (and testing) to switch on the audio when a receiver is operated in a waiting/muted reception mode (see Chapter 4).

M.1.14 Programme TYpe Name (PTYN)

The PTYN feature is used to further describe the current PTY. PTYN permits the display of a more specific PTY description that the broadcaster can freely decide (e.g., PTY = 4: Sport and PTYN: Football). The PTYN is not intended to change the default characters used to display the PTY whose code is used during search or wait modes, but only to show in detail the Programme Type once tuned to a programme. If the broadcaster is satisfied with a default PTY name, it is not necessary to use additional data capacity for PTYN. PTYN is not intended to be used for automatic PTY selection and must not be used for giving sequential information (see Chapter 4).

M.1.15 Radio Paging (RP)

The RP feature of RDS is intended to provide Radio Paging using existing VHF/FM broadcasts as a delivery mechanism, thereby avoiding the need for a dedicated network of transmitters. Subscribers to a paging service require a special paging receiver in which the subscriber address code is stored (see Chapter 8).

M.1.16 RadioText (RT)

The RT feature allows the transmission of text messages up to 64 characters in length to be conveyed primarily to consumer home receivers. Shorter messages are also possible, and they may be frequently altered at a reasonable rate to allow, for example, programme-related information such as music title, conductor, orchestra, and CD reference number in order to decide on the CD to be purchased (see Chapter 4).

M.1.17 Traffic Announcement (TA) Flag

The TA indicator feature is a two-state flag signal to inform a receiver whether there is a traffic announcement on-air or not. This signalling may permit a receiver to switch automatically from any audio mode to the traffic announcement, switch on the traffic announcement automatically when the receiver is in a waiting reception mode and the audio signal is muted, or switch from a programme to another one carrying a traffic announcement (see Chapter 6).

M.1.18 Transparent Data Channels (TDC)

The TDC feature is designed to allow the transmission of free-format nonprogramme-related data in one of 32 channels. TDC was developed for closed user group applications (and is not available in consumer receivers), and special receivers would be required to respond to such data. Now that ODA has been specified, it is not expected to be used for all new applications (see Chapter 5).

M.1.19 Traffic Message Channel (TMC)

The TMC feature was developed to permit coded traffic and travel information to be conveyed to specialised consumer receivers and navigation systems, to give language-*independent* responses/displays when travelling across countries in close proximity to one another. The TMC feature is a special case of an ODA using a predefined group type (8A) and type 3A groups. The TMC system is fully defined in a family of CEN standards in the number sequence: ENV 12313 (see Chapter 7).

M.1.20 Traffic Programme (TP) Flag

The TP indicator feature is a two-state flag signal to inform a receiver whether the transmission being received carries traffic announcements or not. The TP flag must only be set on transmissions that dynamically switch on the TA indi-cator flag *during* traffic announcements. The signal may be used during auto-matic search tuning (see Chapter 3 and 6).

M.2 Additional Definitions

In all areas of broadcasting, new technologies are being introduced. Further-more, deregulation of broadcasting is leading to new relationships among those involved with broadcast services.

This has led to many new functional responsibilities, and the former (lim-ited) definition of the word *broadcaster* needs to be redefined. This word has now acquired colloquially different, assumed meanings according to the state of development in any particular area/country. The key to explaining the functional responsibilities of a broadcaster and associated functions is to select good descriptive word(s) or phrases and define the elements and functions of that word.

The set of words and definitions that follow was originally developed for the RDS situation and has since been considered in the DAB context. Thus, appropriate combinations can be used for any situation, including where RDS and DAB carry the same service. These definitions were adopted by CEN TC 278 SWG 4.1 for use in standards developed therein and have been used in the CEN standard ENV 12313-1.

M.2.1 Broadcaster

A *broadcaster* is a traditionally incorporated organisation responsible for a con-tinuous strand of programmes, their quality, and programme-associated data, as well as being responsible for the overall coordination of "broadcast transmis-sions" (often, a broadcaster is the licensee of a national regulator). A broad-caster is normally a *programme service provider* (of radio and/or television programmes) and may sometimes obtain programmes from another pro-gramme service provider. A broadcaster may also be a data service provider (e.g., RDS-TMC in Germany).

M.2.2 Programme Service Provider

A *programme service provider* is an organisation that manages and originates programmes (radio and/or television programmes) and associated data for broadcast. This will often be carried out by a broadcaster, but allows for the subtle distinction where a separate company is commissioned to produce a programme, together with associated data (e.g., text of teacher notes for an educational series).

M.2.3 Data Service Provider

A *data service provider* is an organisation that manages any data service by gathering data, processing that data, and marketing and providing the data service. A data service provider then negotiates for the use of the necessary data bandwidth with a broadcaster and/or transmission operator. A data service provider is responsible for the "quality" of data to his or her customers and must provide suitable customer support. Editorial control over the data may be part of a service agreement between data service provider and broadcaster. (For example, a TMC service may require the broadcaster to apply some editorial control so that both TMC messages and other broadcast services such as spoken or teletext traffic and travel information—possibly derived from more than one source—are not contradictory.) It is recognised that when talking about a particular service, the word *data* is often dropped; however, this could lead to confusion with a programme service provider, so care should be exercised to ensure the context is well understood.

M.2.4 Transmission Operator

A *transmission operator* is an organisation responsible for the *actual* transmission of the full broadcast signal, including the audio programme, programme-associated data, and data services. Normally, a transmission operator is contracted by a broadcaster to perform the transmission task.

M.2.5 Network Operator

A *network operator* is an organisation contracted to supply both programme and data circuits interconnecting data service provider, programme service provider, broadcaster, and transmission operator. According to the connections, various protocols may be used (e.g., EBU UECP, Eureka 147 ETI and STI, and TPEG).

M.3 Abbreviations

Many abbreviations are commonly used in the context of RDS, and these are listed in Table M.1 in alphabetical order. This includes RDS features and associated components (e.g., specialised items such as the Linkage Set Number). Most of these abbreviations are explained in Section M.1.

Table M.1
RDS Features and Associated Abbreviations

AID	Application IDentification (for ODA)
AF	Alternative Frequencies list
CI	Country Identifier/country code
CT	Clock Time (and date)
DI	Decoder Identification
ECC	Extended Country Code
EG	Extended Generic indicator
EON	Enhanced Other Networks information
EWS	Emergency Warning System
IH	In-House applications
ILS	International Linkage Set indicator
LA	Linkage Actuator
LI	Linkage Identifier
LSN	Linkage Set Number
MS	Music Speech identifier
ODA	Open Data Applications
PI	Programme Identification
PIN	Programme Item Number
PS	Programme Service name
PTY	Programme TYpe
PTYI	Dynamic Programme TYpe Indicator
PTYN	Programme TYpe Name
RBDS	Radio Broadcast Data System
RDS	Radio Data System
RP	Radio Paging
RT	RadioText
TA	Traffic Announcement flag
TDC	Transparent Data Channels
TMC	Traffic Message Channel
TP	Traffic Programme flag

The RDS capability to convey Radio Paging has introduced a number of specialised abbreviations used in connection with the basic paging (BP) and extended paging protocols (EPP), and these are listed in Table M.2. Operation of a Radio Paging service also requires certain other RDS features to be implemented (e.g., CT) and their abbreviations are listed in Table M.1.

The RDS-TMC specification ENV 12313-1 introduces further concepts that are specific to RDS-TMC and their abbreviations are listed in Table M.3.

M.4 Acronyms

In the field of new radio data transmission technologies, and particularly of RDS-TMC, many acronyms have been used to describe functionality,

Table M.2
Radio Paging Abbreviations

BPP	Basic paging protocol
CCF	Current carrier frequency
CS	Cycle selection
EPP	Extended paging protocol
IT	Interval numbering
NI	National international id
OPC	Operator code
PAC	Paging area code
RP	Radio Paging
SI	System information
STY	Subtype group

Table M.3
TMC Abbreviations

AFI	Alternative frequency indicator
LTN	Location table number
MGS	Message geographical scope
SID	Service identifier
TMC	Traffic Message Channel

development projects, and specifications. A selection of those most commonly encountered is shown in Table M.4.

Table M.4
Acronyms Commonly Encountered

Acronym	Description
AA	Automobile Association (U.K.)
ADAC	Allgemeiner Deutscher Automobil Club (German automobile users' association)
ADR	ASTRA Digital Radio
AHS	Automated Highway Systems
ALERT	Advice and Problem Location for European Road Transport (pre 1997 meaning)
ALERT	Agreed layer for European RDS-TMC (introduced July 1997 for EC service MoU)
ALERT-C	Advice and Problem Location for European Road Transport, version C
ALERT-Plus	Advice and Problem Location for European Road Transport, version Plus
AM	Amplitude modulation
AMDS	AM Data System (a baseband coding specification from the EBU for data broadcasting on LF, MF, and HF)
ARI	Autofahrer Rundfunk Information (drivers radio information, used in Austria, Germany, Luxemburg, and Switzerland)
ARD	Arbeitsgemeinschaft der öffentlich rechtlichen Rundfunkanstalten der Bundesrepublik Deutschland (the German public broadcasters' association)
ARTS	Advanced Road Transportation Systems (Japan)
	A European regional project (EC/DG VII)
ATT	Advanced transport telematics (EC)
ATT-ALERT	Advanced Transport Telematics-Advice and Problem Location for European Road Traffic ATT(DRIVE) Project V2028
AVICS	Advanced Vehicle Control & Infrastructure Systems
BBC	British Broadcasting Corporation (the U.K. public broadcaster)
BEVEI	BEssere Verkehrs Information (Better Traffic Information—an ARD (WDR and SWF) and partners RDS-TMC pilot)
BR	Bayerischer Rundfunk (German public radio broadcaster)
CARMINAT	French consortium originally established as a Eureka research project
CEM	Consumer electronics manufacturer
CEMA	Consumer Electronics Manufacturers Association (a branch of EIA in the United States)
CEMT	Conférence Européene des Ministres des Transports (see also ECMT)

Table M.4 (continued)

Acronym	Description
CEN	Comité Européen de Normalisation (responsible for RDS-TMC standards)
CENELEC	Comité Européen de Normalisation Electrotechnique (responsible for the RDS standard)
CENTRICO	A European regional project (EC/DG VII)
CES	Consumers Electronics Show (United States)
CORD	CORD is a strategic assessment of ATT implementation through ATT(DRIVE) Project V2056
CORVETTE	A European regional project (EC/DG VII)
CRC	Cyclic redundancy check
DAB	Digital audio broadcasting, also called digital radio
DARC	Data radio channel (a system for high-speed data broadcasting in FM radio)
DATEX	An initiative to coordinate data protocols used between TICs (subject of EC Service MoU)
DEFI	Definition of first phase implementation (Pan-European RDS-TMC road information service)
DG VII	Directorate General VII of the European Commission (concerned with transport)
DG XIII	Directorate General XIII of the European Commission (concerned with Telematics)
DGPS	Differential Global Positioning Satellite/System (see also GPS)
DRIVE	Dedicated Road Infrastructure for Vehicle safety in Europe
DSP	Data service provider
DVB	Digital video broadcasting
EACEM	European Association of Consumer Electronics Manufacturers
EBU	European Broadcasting Union (see also UER)
EC	European Commission
ECMT	European Conference of Ministers of Transport (see also CEMT)
ECORTIS	A joint project with FORCE to develop final solutions for RDS-TMC implementations
ECU	European currency unit (see also MECU)
EDIFACT	Electronic Data Interchange For Administration, Commerce, and Trade
EDRM	European digital road map
EIA	Electronics Industry Association (United States)
EN	European standard (see prEN)
ENV	European prestandard (see prENV)

Table M.4 (continued)

Acronym	Description
EPISODE	European Preoperational Implementation Survey On further Development and Evaluation of RDS-TMC (broadcast sector)
ERTICO	European Road Transport Telematics Implementation Coordination Organisation
ETS	European telecommunications standard from ETSI
ETSI	European Telecommunications Standards Institute
EU	European Union
EUREKA	European Research Coordination Agency
EUROAD	Location coding concept in RDS-TMC for the provision of transborder event messages
FM	Frequency modulation
FORCE	A joint project with ECORTIS to develop final technology and solutions for RDS-TMC implementations
GDF	Geographical data file
GEMINI	Generation of Event Messages in the New Integrated Road Transport Environment through the ATT(DRIVE) Project V2038
GIS	Geo-information systems
GPS	Global Positioning Satellite/System (see also DGPS)
GSM	Groupe Systeme Mobile (standard for worldwide digital cell phone technology)
HMI	Human/machine interface (see also MMI)
HSDS	High-speed data system (FM broadcasting)
IEC	International Electrotechnical Committee
IFA	Internationale Funkausstellung (International fair of audiovisual consumer products)
IRT	Institut für Rundfunktechnik (German public broadcasters' technology research centre)
IRTE	Integrated road transport environment
ISO	International Standards Organisation
ITS	Intelligent transport systems (EC)
ITT	Intelligent transport telematics (EC)
ITU	International Telecommunication Union
IVHS	Intelligent Vehicle Highway System (United States)
LMSK	Level controlled modulation shift keying
MECU	Million European currency units (see also ECU)
MMI	Man machine interface (see also HMI)

Table M.4 (continued)

Acronym	Description
MoU	Memorandum of Understanding (see also ALERT and DATEX)
MS	Member states (of European union)
NAB	National Association of Broadcasters (United States)
NDR	Norddeutscher Rundfunk (German public radio broadcaster)
NOB	Nederlands Omroepproduktie Bedrijf nv (Netherlands public broadcast production company)
NOS	Nederlandse Omroep Stichting (Netherlands public radio broadcaster)
NRSC	National Radio Systems Committee of NAB/EIA (United States)
OSI	Open systems interconnection
ORF	Österreichischer Rundfunk – Austrian public radio broadcaster)
PES	Pan-European service
PLEIADES	Paris London Corridor Evaluation of Integrated ATT and DRIVE Experimental Systems through a ATT(DRIVE) Project V2047
prEN	Draft European standard (see also EN)
prENV	Draft European prestandard (see also ENV)
PSP	Programme service provider
RAI	Radiotelevisione Italiana (Italian public radio broadcaster)
RASANT	Radio Aided Satellite Navigation Technique (a German DGPS system for FM radio)
RF	Radio France (French public radio broadcaster)
RPA	Radio Paging Association (operated within the RDS Forum)
RTCC	Regional Traffic Control Centre
RTCM	Radio Technical Commission for Maritime Services
RTD	Research and technology demonstration
RTI	Road traffic informatics
RTT	Road transport telematics
RVC	Road vehicle communications (an ECMT subgroup)
SDR	Süddeutscher Rundfunk (German public radio broadcaster)
SERTI	A European regional project (EC/DG VII)
SME	Small- and medium-sized enterprise
SR	Sveriges Radio (the Swedish public radio broadcaster)
SWF	Südwestfunk (German public broadcaster)
SWG	Sub-working group (refer to CEN or CENELEC)
TAP	Telematics applications programme

Table M.4 (continued)

Acronym	Description
TC	Technical committee (refer to CEN or CENELEC)
TCC	Traffic Control Centre
TDF	Télédiffusion de France (a transmission operator that is part of France Telecom group)
TELTEN	TELematic implementation on the Trans European Road Network (DG VII project)
TEN	Trans-European network
TPEG	Transport programme experts group specification developed by the EBU
TERN	Trans-European Road Network
TIC	Traffic Information Centre (in some countries this only covers road information)
TICS	Traffic Information Centre Systems (refer to CEN TC204)
TTI	Traffic and travel information
TTIC	Traffic and Travel Information Centre
UECP	Universal Encoder Communication Protocol (specification from the EBU)
UER	Union Européenne de Radio-Télévision (see also EBU)
VDA	Verband Deutscher Automobilhersteller (German automobile industry association)
VHF/FM	Very high frequency/frequency modulation (used in Band II, 87.5-108 MHz)
VBI	Vertical blanking interval (in analogue television that makes it possible to accommodate data)
VICS	Vehicle Information & Communication System
VIKING	A European regional project (EC/DG VII)
VMS	Variable message signs
WAP	Wireless application protocol
WDR	Westdeutscher Rundfunk (German public broadcaster)
WG	Working group (refer to CEN and CENELEC)

References

[1] EN 50067:1998 - Specification of the Radio Data System (RDS) for VHF/FM sound broadcasting in the frequency range from 87.5 to 108.0 MHz.

About the Authors

Dietmar Kopitz has coordinated all RDS development work within the European Broadcasting Union since the 1970s, participating in specification writing, elaboration of the implementation guidelines, and the European and U.S. standardisation processes. He is a founder and co-chairman of the RDS Forum. In the field of RDS-TMC, he led the European Commission funded EPISODE project.

Bev Marks contributed to the initial BBC RDS studio and transmitter networks implementations. For several years he has been a freelance broadcast engineer working on a number of special projects for the EBU related to radio data broadcasting. He is involved in the RDS Forum on RDS guidelines development and in the EPISODE project concerning standardisation of RDS-TMC.

Index

Recent Titles in the Artech House Mobile Communications Series

John Walker, Series Editor

For further information on these and other Artech House titles, including previously considered out-of-print books now available through our In-Print-Forever® (IPF®) program, contact:

Artech House
685 Canton Street
Norwood, MA 02062
Phone: 781-769-9750
Fax: 781-769-6334
e-mail: artech@artechhouse.com

Artech House
46 Gillingham Street
London SW1V 1AH UK
Phone: +44 (0)171-973-8077
Fax: +44 (0)171-630-0166
e-mail: artech-uk@artechhouse.com

Find us on the World Wide Web at:
www.artechhouse.com